MW00843399

Information Theory Tools for Visualization

A K PETERS VISUALIZATION SERIES

Series Editor: Tamara Munzner

Information Theory Tools for Visualization

Min Chen, Miquel Feixas, Ivan Viola, Anton Bardera, Han-Wei Shen, and Mateu Sbert

Visualization Analysis and Design

Tamara Munzner

Information Theory Tools for Visualization

Min Chen
University of Oxford
United Kingdom

Miquel Feixas
University of Girona
Spain

Ivan Viola
TU Wien
Austria

Anton Bardera
University of Girona
Spain

Han-Wei Shen
The Ohio State University
Columbus, USA

Mateu Sbert
University of Girona
Spain

Tianjin University
China

CRC Press
Taylor & Francis Group
Boca Raton London New York

CRC Press is an imprint of the
Taylor & Francis Group, an **informa** business

AN A K PETERS BOOK

CRC Press
Taylor & Francis Group
6000 Broken Sound Parkway NW, Suite 300
Boca Raton, FL 33487-2742

© 2017 by Taylor & Francis Group, LLC
CRC Press is an imprint of Taylor & Francis Group, an Informa business

No claim to original U.S. Government works

Printed on acid-free paper
Version Date: 20160502

International Standard Book Number-13: 978-1-4987-4093-7 (Hardback)

This book contains information obtained from authentic and highly regarded sources. Reasonable efforts have been made to publish reliable data and information, but the author and publisher cannot assume responsibility for the validity of all materials or the consequences of their use. The authors and publishers have attempted to trace the copyright holders of all material reproduced in this publication and apologize to copyright holders if permission to publish in this form has not been obtained. If any copyright material has not been acknowledged please write and let us know so we may rectify in any future reprint.

Except as permitted under U.S. Copyright Law, no part of this book may be reprinted, reproduced, transmitted, or utilized in any form by any electronic, mechanical, or other means, now known or hereafter invented, including photocopying, microfilming, and recording, or in any information storage or retrieval system, without written permission from the publishers.

For permission to photocopy or use material electronically from this work, please access www.copyright. com (http://www.copyright.com/) or contact the Copyright Clearance Center, Inc. (CCC), 222 Rosewood Drive, Danvers, MA 01923, 978-750-8400. CCC is a not-for-profit organization that provides licenses and registration for a variety of users. For organizations that have been granted a photocopy license by the CCC, a separate system of payment has been arranged.

Trademark Notice: Product or corporate names may be trademarks or registered trademarks, and are used only for identification and explanation without intent to infringe.

Library of Congress Cataloging-in-Publication Data

Names: Chen, Min, 1960 May 25- author.
Title: Information theory tools for visualization / authors, Min Chen, Miquel
Feixas, Ivan Viola, Anton Bardera, Han-Wei Shen, and Mateu Sbert.
Description: Boca Raton : Taylor & Francis, CRC Press, 2017. | Series: A K
Peters visualization series | Includes bibliographical references and
index.
Identifiers: LCCN 2016018492 | ISBN 9781498740937 (alk. paper)
Subjects: LCSH: Information visualization. | Information theory.
Classification: LCC QA76.9.I52 C45 2017 | DDC 001.4/226--dc23
LC record available at https://lccn.loc.gov/2016018492

Visit the Taylor & Francis Web site at
http://www.taylorandfrancis.com

and the CRC Press Web site at
http://www.crcpress.com

Printed and bound in the United States of America by
Edwards Brothers Malloy on sustainably sourced paper

Contents

Foreword

On this date 100 years ago[1], Claude Shannon was born in Petoskey, Michigan, USA. After graduating from MIT in 1940, Shannon joined Bell Labs, where in 1948 he would produce a two-part memorandum called "A Mathematical Theory of Communication," which focused on the problem of how best to encode the information a sender wants to transmit. In this memorandum, Shannon developed information entropy as a measure for uncertainty in a message, which helped to define the new field of information theory.

Fast-forward a century, and we have an exciting new book, *Information Theory Tools for Visualization*, which applies information theory to the field of visual analysis, Each of its six authors, Min Chen, Miquel Feixas, Ivan Viola, Anton Bardera, Han-Wei Shen, and Mateu Sbert, has made significant contributions to the nascent field of information theory in visualization. In fact, the authors represent the avant-garde in the new field of visualization theory, having produced among them over 30 papers and referenced articles on the topic before and after the turn of the millennium. One will notice that the vast number of papers published by the authors and referenced articles on the topic of information theory in visualization have been published after the year 2000 (most even more recently). In this book, then, a reader will be guided into an exciting new field by the very people who are creating it and generating its new research results.

Information Theory Tools for Visualization begins by introducing basic concepts in information theory. The book then moves to create a foundation and metrics for applying information theory in visualization, using analogies with other fields to make links between information theory and visualization. In Chapters 3, 4, and 5, the authors apply information theory to different aspects of scientific visualization, specifically investigating information theory applications in volume visualization and flow visualization, as well as using information theory to define optimal viewpoint metrics for surface and volume visualization applications. The book concludes with a chapter on applying information theory to information visualization, exploring topics on parallel coordinates, trees, graphs, and multivariate data.

I hope you will enjoy following the authors as they lead you into the new field of information theory in visualization.

---

[1] During this centennial year of Shannon's birth, there are a number of wonderful articles about Shannon and his work in information theory, including these from the IEEE Information Theory Society: http://www.itsoc.org/resources/Shannon-Centenary

Sincerely,

Chris Johnson, PhD
Director,
   Scientific Computing and Imaging Institute (www.sci.utah.edu)
Distinguished Professor,
   School of Computing
   University of Utah

Preface

This book represents the first serious attempt to bring together all major uses of information theory in the visualization literature up to now. We, the authors of this book, started this journey in the autumn of 2014 with our individual branches of knowledge about the existing works. By the end of summer 2015, when the first draft was compiled from separately written chapters, we were astonished by the coverage and diversity of the footprints of information theory in the landscape of visualization. We hope that readers of this work will share our enthusiasm and confidence about the promising role of information theory in visualization. While information theory has provided volume visualization, flow visualization, and information with a variety of techniques, it has also shown its potential to underpin the theoretical foundation of visualization. Of course, we are fully aware that this is only the beginning. We therefore hope that this book will inspire a surge of new effort in furthering advancement in this endeavour.

The book is organized to provide easy access to the basics of information theory while facilitating in-depth treatments of major knowledge domains for readers who follow a particular research agenda.

Chapter 1 introduces the basic concepts in information theory. Some of these concepts, such as *entropy* and *mutual information*, are prerequisites for comprehending all of the following chapters. Others allow readers to extend their knowledge of information theory in preparation for more in-depth discussions in some chapters as well as for the development of new research projects.

Chapter 2 presents an information-theoretic framework for visualization by establishing links between visualization processes and data communications, suggesting the potential of using information theory to underpin the theoretic foundation of visualization. Examples of various phenomena in visualization processes, including visual perception, visual mapping, interaction, and process optimization, are used to demonstrate the fundamental role of information theory in visualization.

Chapter 3 focuses on the early application of information theory in visualization and computer graphics. Given any 3D scene featuring occlusion or translucency, we are looking for the best view. As the quality of a view depends on the declared objective, and can be considered as an information transfer, different measures are proposed that fit within the framework of an information channel between viewpoints and voxels. These information-theoretic metrics can be used to support viewpoint optimization, focus of attention, and illustrative rendering.

Chapter 4 builds on Chapter 3 and gives an in-depth treatment of the uses of information theory in volume visualization. The footprints of information-theoretic techniques traverse from time-varying data to multimodal data, from isosurface extraction to transfer function design, and from level-of-detail organization to volume splitting.

Chapter 5 gives an in-depth treatment of the uses of information theory in flow visualization. It begins with an extension of the basic concepts in Chapter 1 to vector fields, which are typically featured in the source data of flow visualization. It then presents information-theoretic techniques for measuring the complexity of streamlines, and for selecting and placing seeds in streamline generation.

Chapter 6 gives an in-depth treatment of the uses of information theory in information visualization. The applications of information theory include quality metrics for guiding multivariate data exploration and generating parallel coordinates plots, summary trees, and privacy-preserving visualization. The chapter also includes a representation for visualizing information-theoretic measurements used in studying multiple variables.

The intended readership of this book includes visualization researchers, undergraduate and postgraduate students in various programs featuring information visualization, scientific visualization, and visual analytics. The book does not require readers to have any prior knowledge of information theory, though some basic knowledge of probability and statistics will be helpful in grasping Chapter 1. At the same time, the book does not require readers to have extensive experience in designing or implementing visualization systems, though some appreciation of the purposes of visualization and its deployments in a broad range of applications will be beneficial.

The application of information theory in visualization is still very much in its infancy. We hope that you will enjoy the book, and perhaps take on the challenge to make further advancements in this area.

ACKNOWLEDGMENTS

Min Chen would like to thank two outstanding research collaborators, Professor Dr. Heike Leitte (née Jänicke), Technische Universität Kaiserslautern, and Professor Amos Golan, American University, Washington, DC, who contributed to the joint works on the information-theoretic framework for visualization and cost–benefit analysis of data analysis and visualization processes, respectively. He would also like to thank his colleagues at Oxford who contributed to the works on quasi-Hamming distance and visual multiplexing: Dr. Phil A. Legg, Dr. Eamonn Maguire, Dr. Simon Walton, Dr. Kai Berger, Dr. Jeyarajan Thiyagalingam, Dr. Brian Duffy, Dr. Hui Fang, Dr. Cameron Holloway, and Professor Anne E. Trefethen. Between 2009 and 2015, the in-depth technical discussions with many of his colleagues provided practical evidence for the information-theoretic works described in Chapter 2. These colleagues include Professor Luciano Floridi (Oxford), Dr. Rita Borgo (Swansea), Dr. Phil W. Grant (Swansea), Dr. Robert S. Laramee (Swansea), Professor Peter Townsend (Swansea), Dr. Gary K. L. Tam (Swansea), Dr. Alfie Abdul-Rahman (Oxford), Dr. David H. S. Chung (Swansea), Dr. Karl J. Proctor (Swansea and Oxford), Dr. Irene Reppa (Swansea), Dr. Farhan Mohamed (Swansea), Saiful Khan (Oxford), and Matthew L. Parry (Swansea). In addition, Min Chen is grateful for the opportunity to conduct joint work on privacy-preserving visualization (featured in Chapter 6) with Dr. Aritra Dasgupta and Dr. Robert Kosara (then University of North Carolina, Charlotte) and on anomaly detection with Professor David Ebert, Sungahn Ko, and their colleagues at Purdue University). Many of the works described in Chapter 2 were carried out with financial support from EPSRC (EP/G006555 and EP/J020435), United Kingdom and United States governmental agencies, and the Info-Metric Institute.

Miquel Feixas, Anton Bardera, and Mateu Sbert would like to acknowledge support from grants TIN2013-47276-C6-1-R from the Spanish government, 2014 SGR 1232 from the Catalan government, and Nos. 61372190, 61331018, and 61571439 from the National Natural Science Foundation of China. Mateu Sbert acknowledges support from 1000 Plan from Tianjin municipality, China. They wish to thank coauthors of information theory visualization papers Imma Boada, Marc Ruiz, Roger Bramon, Quim Rodríguez, Josep Puig, Stefan Bruckner, and Meister Edi Gröller.

Ivan Viola would like to thank his colleagues from TU Wien, Austria, and University of Bergen, Norway, most notably Meister Edi Gröller, Stefan Bruckner, and Helwig Hauser, for doing exciting research together. Ivan Viola has been funded by the Vienna Science and Technology Fund (WWTF) through project VRG11-010 and additionally supported by the EC Marie Curie Career Integration Grant through project PCIG13-GA-2013-618680.

Han-Wei Shen would like to thank his current and former students Teng-Yok Lee, Ayan Biswas, Lijie Xu, Soumya Dutta, Wenbin He, Chun-Ming Chen, Xiaotong Liu, Tzu-Hsuan Wei, Xin Tong, Kewei Lu, Ko-Chi Wang, Subhashis Hazarika, Li Cheng, Udeepta Bordoloi, and Chaoli Wang. Han-Wei Shen's research has been primarily supported by the US Department of Energy and the National Science Foundation.

The authors would like to thank the series editor, Professor Tamara Munzner, University of British Columbia. The authors are most grateful for the support received from the staff at Taylor & Francis, including Sunil Nair, Karen Simon, Amber Donley, Alex Edwards, Kevin Craig, Shashi Kumar, and Marcus Fontaine. Without their skillful project management, copy-editing, cover design, and various advice, our journey would have taken much longer and encountered many more difficulties.

CHAPTER 1

Basic Concepts of Information Theory

CONTENTS

Information theory was founded in 1948 and described in Shannon's paper "A Mathematical Theory of Communication" [107]. In this paper, Shannon defined entropy and mutual information (initially called the *rate of transmission*), and introduced the fundamental laws of data compression and transmission. In information theory, *information* is simply the outcome of a selection among a finite number of possibilities and an *information source* is modeled as a random variable or as a random process. While Shannon entropy expresses the uncertainty or the information content of a single random variable, mutual information quantifies the dependence between two random variables and plays an important role in the analysis of a *communication channel*, a system in which the output depends probabilistically on its input [33, 134, 148]. From its birth to date, information theory has interacted with many different fields, such as statistical inference, computer science, mathematics, physics, chemistry, economics, and biology.

This chapter presents Shannon's information measures (entropy, conditional entropy, and mutual information) for discrete and continuous random variables, together with Kullback–Leibler distance, entropy rate, some relevant inequalities, and the information bottleneck method. Two main references on information theory are the books by Cover and Thomas [33] and Yeung [148].

1.1 ENTROPY

Let X be a discrete random variable with alphabet $\mathbb{X}$ and probability distribution $\{p(x)\}$, where $p(x) = \Pr\{X = x\}$ and $x \in \mathbb{X}$. The probability distribution $\{p(x)\}$ will also be denoted by $p(X)$ or simply p. For instance, a discrete random variable can be used to describe the toss of a fair coin, with alphabet $\mathbb{X} = \{head, tail\}$ and probability distribution $p(X) = \{1/2, 1/2\}$. The *entropy* $H(X)$ of a discrete random variable X is defined by

$$H(X) = -\sum_{x \in \mathbb{X}} p(x) \log p(x), \tag{1.1}$$

where the summation is over the corresponding alphabet. Logarithms are taken in base 2 and, as a consequence, entropy is expressed in bits. The convention $0 \log 0 = 0$ is justified by continuity since $x \log x \to 0$ as $x \to 0$.

As the term $-\log p(x)$ is interpreted as the information content (or uncertainty) associated with the result x, the entropy represents the average amount of information (or uncertainty) of a random variable. Entropy is denoted by $H(X)$ or $H(p)$, where p stands for the probability distribution $p(X)$.

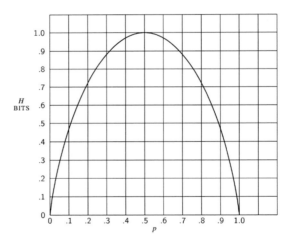

Figure 1.1 Plot of binary entropy used by Shannon in [107].

The binary entropy (Figure 1.1) of a random variable X with alphabet $\{x_1, x_2\}$ and probability distribution $\{p, 1 - p\}$ is given by

$$H(X) = -p \log p - (1 - p) \log(1 - p). \tag{1.2}$$

The maximum binary entropy is 1 bit when $p = 1/2$. Thus, the entropy of a fair coin toss is given by $H(X) = -(1/2) \log(1/2) - (1/2) \log(1/2) = \log 2 = 1$ bit. For the toss of a fair die with alphabet $\mathbb{X} = \{1, 2, 3, 4, 5, 6\}$ and probability distribution $p(X) = \{1/6, 1/6, 1/6, 1/6, 1/6, 1/6\}$, $H(X) = \log 6 = 2.58$ bits.

Some relevant properties of entropy [107] are as follows:

1. $0 \leq H(X) \leq \log |\mathbb{X}|$

   (a) $H(X) = 0$ when all the probabilities are zero except one with unit value.

   (b) $H(X) = \log |\mathbb{X}|$ when all the probabilities are equal.

2. If the probabilities are equalized, entropy increases.

3. If an event is broken down into two successive events, the original H should be the weighted sum of the individual values of H. The meaning of this property, called the *grouping property*, is illustrated in Figure 1.2. On the left, we have three possibilities with probabilities $p_1 = 1/2$, $p_2 = 1/3$, $p_3 = 1/6$. On the right, we first choose between two possibilities, each with probability $1/2$, and if the second occurs, we make another choice with probabilities $2/3$, $1/3$. The final results have the same probabilities as before. In this example, it is required that $H(1/2, 1/3, 1/6) = H(1/2, 1/2) + (1/2)H(2/3, 1/3)$. The coefficient $1/2$ is because the second choice occurs with this probability.

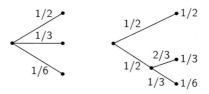

Figure 1.2 Example of the grouping property of the entropy used by Shannon in [107].

For a pair of discrete random variables X and Y with a joint probability distribution $p(X,Y) = \{p(x,y)\}$, the *joint entropy* $H(X,Y)$ is defined by

$$H(X,Y) = -\sum_{x \in \mathbb{X}} \sum_{y \in \mathbb{Y}} p(x,y) \log p(x,y), \tag{1.3}$$

where $p(x,y) = \Pr[X = x, Y = y]$ is the joint probability of x and y.

The *conditional entropy* $H(Y|X)$ of a random variable Y given a random variable X is defined by

$$
\begin{aligned}
H(Y|X) &= \sum_{x \in \mathbb{X}} p(x) H(Y|X = x) \\
&= \sum_{x \in \mathbb{X}} p(x) \left(-\sum_{y \in \mathbb{Y}} p(y|x) \log p(y|x) \right) \tag{1.4} \\
&= -\sum_{x \in \mathbb{X}} \sum_{y \in \mathbb{Y}} p(x,y) \log p(y|x), \tag{1.5}
\end{aligned}
$$

where $p(y|x) = \Pr[Y = y|X = x]$ is the conditional probability of y given x.

The Bayes theorem expresses the relationship between marginal probabilities $p(x)$ and $p(y)$, conditional probabilities $p(y|x)$ and $p(x|y)$, and joint probabilities $p(x, y)$:

$$p(x, y) = p(x)p(y|x) = p(y)p(x|y). \tag{1.6}$$

If X and Y are independent, then $p(x, y) = p(x)p(y)$. Marginal probabilities can be obtained from $p(x, y)$ by summation: $p(x) = \sum_{y \in \mathbb{Y}} p(x, y)$ and $p(y) = \sum_{x \in \mathbb{X}} p(x, y)$. The conditional probability distribution of Y given x is denoted by $p(Y|x)$ and the transition probability matrix (i.e., the matrix whose rows are given by $p(Y|x)$) is denoted by $p(Y|X)$.

Equation 1.4 expresses the conditional entropy of Y given X and is defined as the expected value of the entropies of the conditional distributions $p(Y|x)$. The conditional entropy can be thought of in terms of a communication channel or *information channel* $X \to Y$ whose output Y depends probabilistically on its input X. This information channel is characterised by a transition probability matrix, which determines the conditional distribution of the output given the input [33]. Hence, $H(Y|X)$ corresponds to the uncertainty in the channel output from the sender's point of view, and vice versa for $H(X|Y)$. Note that in general $H(Y|X) \neq H(X|Y)$.

The following properties are fulfilled:

1. $H(X, Y) = H(X) + H(Y|X) = H(Y) + H(X|Y)$.

2. $H(X, Y) \leq H(X) + H(Y)$.

3. $H(X) \geq H(X|Y) \geq 0$.

4. If X and Y are independent, then $H(Y|X) = H(Y)$ since $p(y|x) = p(y)$ and, consequently, $H(X, Y) = H(X) + H(Y)$.

5. If $p(X) = p(Y)$, then $H(Y|X) = H(X|Y)$.

1.2 RELATIVE ENTROPY AND MUTUAL INFORMATION

Given two probability distributions p and q, defined on a common alphabet, the relative entropy quantifies how much p is different from q. The *relative entropy* or *Kullback–Leibler distance* $D_{KL}(p||q)$ between two probability distributions p and q defined over the alphabet $\mathbb{X}$ is given by

$$D_{KL}(p||q) = \sum_{x \in \mathbb{X}} p(x) \log \frac{p(x)}{q(x)}. \tag{1.7}$$

The conventions that $0 \log(0/0) = 0$ and $a \log(a/0) = \infty$ if $a > 0$ are adopted. The relative entropy satisfies the divergence inequality or information inequality

$$D_{KL}(p||q) \geq 0, \tag{1.8}$$

with equality if and only if $p = q$. The relative entropy is also called *informa-tion divergence* [36] or *informational divergence* [148]. It is not strictly a metric since it does not satisfy both the property of symmetry, $d(x,y) = d(y,x)$, and the triangle inequality, $d(x,y) + d(y,z) \geq d(x,z)$, where d represents a metric function.

The *mutual information* $I(X;Y)$ between two random variables X and Y is defined by

$$I(X;Y) = H(X) - H(X|Y) = H(Y) - H(Y|X) \qquad (1.9)$$

$$= \sum_{x \in \mathbb{X}} \sum_{y \in \mathbb{Y}} p(x,y) \log \frac{p(x,y)}{p(x)p(y)} \qquad (1.10)$$

$$= \sum_{x \in \mathbb{X}} p(x) \sum_{y \in \mathbb{Y}} p(y|x) \log \frac{p(y|x)}{p(y)}. \qquad (1.11)$$

Mutual information (MI) represents the amount of information that one random variable, the input of the channel, contains about a second random variable, the output of the channel, and vice versa. That is, mutual information expresses how much the knowledge of Y decreases the uncertainty of X, and vice versa. $I(X;Y)$ is a measure of the shared information or dependence between X and Y. Thus, if X and Y are independent, then $I(X;Y) = 0$.

Mutual information is a special case of relative entropy since it can be expressed as the Kullback–Leibler distance between the joint distribution and the product of marginal distributions:

$$I(X;Y) = D_{KL}(p(X,Y)||p(X)p(Y)). \qquad (1.12)$$

Mutual information $I(X;Y)$ fulfills the following properties:

1. $I(X;Y) \geq 0$ with equality if and only if X and Y are independent.

2. $I(X;Y)$ is symmetrical in X and Y: $I(X;Y) = I(Y;X)$.

3. $I(X;X) = H(X)$.

4. $I(X;Y) = H(X) + H(Y) - H(X,Y)$.

The relations between Shannon's information measures are summarized by the information diagram of Figure 1.3, which is a variation of the Venn diagram because the universal set is not represented. Observe that $I(X;Y)$ and $H(X,Y)$ are represented, respectively, by the intersection and the union of the information in X (represented by $H(X)$) with the information in Y (represented by $H(Y)$). $H(X|Y)$ is represented by the difference between the information in X and the information in Y, and vice versa for $H(Y|X)$. The correspondence between Shannon's information measures and set theory is discussed by Yeung [148].

The following inequality, called the *data processing inequality*, is defined for

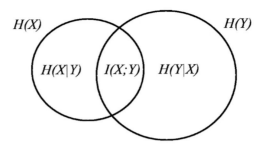

Figure 1.3 The information diagram shows the relationship between Shannon's information measures.

variables that form a Markov chain. The variables X, Y, and Z form a Markov chain $X \to Y \to Z$ if $p(x, y, z) = p(x)p(y|x)p(z|y)$. That is, the probability of the future state depends on the current state only and is independent of what happened before the current state. A more general definition of a Markov chain is given in Section 1.4.

The *data processing inequality* is expressed as follows. If $X \to Y \to Z$ is a Markov chain, then

$$I(X;Y) \geq I(X;Z). \tag{1.13}$$

This result proves that no processing of Y, deterministic or random, can increase the information that Y contains about X. In particular, if $Z = f(Y)$, then $X \to Y \to f(Y)$ and, consequently, $I(X;Y) \geq I(X; f(Y))$ [33].

1.3 INFORMATION SPECIFIC TO A PARTICULAR SYMBOL

As we have seen in the previous section, mutual information is a measure of the correlation or dependence between two random variables. In this section, mutual information $I(X;Y)$ is decomposed in different ways to obtain the specific information associated with a symbol of an alphabet $\mathbb{X}$ (or $\mathbb{Y}$). In the field of neural systems, several definitions of specific information have been proposed to quantify the information or relevance associated with stimuli and responses [15, 41].

For random variables X and Y, representing an ensemble of stimuli (or input values) $\mathbb{X}$ and a set of responses (or output values) $\mathbb{Y}$, respectively, mutual information (see Equations 1.9 and 1.11) is given by

$$I(X;Y) = H(Y) - H(Y|X) \tag{1.14}$$

$$= \sum_{x \in \mathbb{X}} p(x) \sum_{y \in \mathbb{Y}} p(y|x) \log \frac{p(y|x)}{p(y)}. \tag{1.15}$$

To quantify the information associated with each stimulus or response,

$I(X;Y)$ can be decomposed as

$$I(X;Y) \quad = \quad \sum_{x \in X} p(x) I(x;Y) \tag{1.16}$$

$$= \quad \sum_{y \in Y} p(y) I(X;y), \tag{1.17}$$

where $I(x;Y)$ and $I(X;y)$ represent, respectively, the information associated with stimulus x and response y. Thus, $I(X;Y)$ can be seen as a weighted average over individual contributions from particular stimuli or particular responses. There exists an infinity of alternatives to define the contribution $I(x;Y)$ or $I(X;y)$, but we present here the three most basic specific information measures, denoted by I_1, I_2, and I_3 [41, 15], that fulfill Equation 1.16.

The *surprise* I_1 can be directly derived from Equation 1.15, taking the contribution of a single stimulus to $I(X;Y)$:

$$I_1(x;Y) = \sum_{y \in Y} p(y|x) \log \frac{p(y|x)}{p(y)}. \tag{1.18}$$

The measure I_1 expresses the surprise about Y from observing x. It can be shown that I_1 is the only positive decomposition of $I(X;Y)$ [41]. This positivity can be proved from the fact that $I_1(x;Y)$ is the Kullback–Leibler distance (Equation 1.7) between $p(Y|x)$ and $p(Y)$.

The *specific information* I_2 [41] can be derived from Equation 1.14, taking the contribution of a single stimulus to $I(X;Y)$:

$$I_2(x;Y) \quad = \quad H(Y) - H(Y|x) \tag{1.19}$$

$$= \quad -\sum_{r \in Y} p(y) \log p(y) + \sum_{y \in Y} p(y|x) \log p(y|x).$$

The measure I_2 expresses the change in uncertainty about Y when x is observed. Note that I_2 can take negative values. This means that certain observations x do increase our uncertainty about the state of the variable Y. It is important to emphasise that I_2 has the property of additivity. An information measure is additive when the information obtained about X from two observations, $y \in Y$ and $z \in Z$, is equal to that obtained from y plus that obtained from z when y is known. Thus, according to DeWeese and Meister [41], the specific information I_2 fulfils the requirement that the information obtained about a sensory stimulus from observing two neurons in a population equals the information from the first neuron plus the information gained from the second neuron after one had already observed the first.

The *stimulus-specific information* I_3 is defined [15] by

$$I_3(x;Y) = \sum_{y \in Y} p(y|x) I_2(X;y). \tag{1.20}$$

The measure I_3 also fulfills Equation 1.16 (for a proof, see [15]). A large value of $I_3(x; Y)$ means that the states of Y associated with x are very informative in the sense of $I_2(X; y)$. That is, the most informative input values X are those that are related to the most informative output values y. Butts [15] proposes some examples that show how I_3 identifies the most significant stimuli.

Similar to the above definitions for a stimulus x, the information associated with a response y could be defined. The properties of positivity and additivity of these measures have been studied in [15, 41]. While I_1 is always positive and non-additive, I_2 can take negative values but is additive, and I_3 can take negative values and is non-additive.

1.4   ENTROPY RATE

A stochastic process or a discrete-time information source $\{X_i\}$ is an indexed sequence of random variables characterized by the joint probability distribution $p(x_1, x_2, \ldots, x_n) = \Pr\{(X_1, X_2, \ldots, X_n) = (x_1, x_2, \ldots, x_n)\}$ with $(x_1, x_2, \ldots, x_n) \in \mathbb{X}^n$ for $n \geq 1$ [33, 148]. Using the property $H(X_1, X_2) = H(X_1) + H(X_2|X_1)$ (Section 1.1) and the induction on n [148], it can be proved that the joint entropy of a collection of n random variables $X_1, \ldots, X_n$ is given by

$$H(X_1, \ldots, X_n) = H(X_1) + \sum_{i=2}^{n} H(X_i|X_1, \ldots, X_{i-1}). \tag{1.21}$$

We now introduce the entropy rate that quantifies how the entropy of a sequence of n random variables increases with n. The *entropy rate* or *entropy density* h^x of a stochastic process $\{X_i\}$ is defined by

$$h^x = \lim_{n \to \infty} \frac{1}{n} H(X_1, X_2, \ldots, X_n) \tag{1.22}$$

when the limit exists. The entropy rate represents the average information content per symbol in a stochastic process.

A stochastic process $\{X_i\}$ is stationary if two subsets of the sequence, $\{X_1, X_2, \ldots, X_n\}$ and $\{X_{1+l}, X_{2+l}, \ldots, X_{n+l}\}$, have the same joint probability distribution for any $n, l \geq 1$: $\Pr\{(X_1, \ldots, X_n) = (x_1, x_2, \ldots, x_n)\} = \Pr\{(X_{1+l}, X_{2+l}, \ldots, X_{n+l}) = (x_1, x_2, \ldots, x_n)\}$. That is, the statistical properties of the process are invariant to a shift in time. For a stationary stochastic process, the entropy rate exists and is equal to

$$h^x = \lim_{n \to \infty} h^x(n), \tag{1.23}$$

where $h^x(n) = H(X_1, \ldots, X_n) - H(X_1, \ldots, X_{n-1}) = H(X_n|X_{n-1}, \ldots, X_1)$. The entropy rate can be seen as the uncertainty associated with a given symbol if all the preceding symbols are known. The entropy rate of a sequence measures the average amount of information (i.e., irreducible randomness) per symbol x and the optimal achievement for any possible compression algorithm [33, 45].

Inspired by the work of Feldman and Crutchfield [35], we present an alternative notation to define the entropy rate. Given a chain $X_1 X_2 X_3 \ldots$ of random variables X_i taking values in $\mathbb{X}$, a block of L consecutive random variables is denoted by $X^L = X_1 \ldots X_L$. The probability that the particular L-block x^L occurs is denoted by $p(x^L)$. The joint entropy of length-L sequences or L-*block entropy* is now denoted by

$$H(X^L) = - \sum_{x^L \in \mathbb{X}^L} p(x^L) \log p(x^L), \qquad (1.24)$$

where the sum runs over all possible L-blocks. The *entropy rate* can be rewritten as

$$
\begin{aligned}
h^x &= \lim_{L \to \infty} \frac{H(X^L)}{L} = \lim_{L \to \infty} h^x(L) \\
&= \lim_{L \to \infty} (H(X^L) - H(X^{L-1})) \qquad (1.25)
\end{aligned}
$$

where $h^x(L) = H(X_L | X_{L-1}, X_{L-2}, \ldots, X_1)$ is the entropy of a symbol conditioned on a block of $L - 1$ previous symbols [35].

A stochastic process $\{X_i\}$ is a Markov chain if $\Pr(X_{n+1} = x_{n+1} | X_n = x_n, \ldots, X_1 = x_1) = \Pr(X_{n+1} = x_{n+1} | X_n = x_n)$, for $n = 1, 2, \ldots$ and for all $x_i \in \mathbb{X}$. If $\{X_i\}$ is a Markov chain, then X_n is called the state at time n. A stationary Markov chain is characterized by its initial state and a transition probability matrix $\mathbf{P} = \{p_{j|i}\}$, where $p_{j|i} = \Pr\{X_{n+1} = j | X_n = i\}$ is called the *transition probability*. A distribution on the states such that the distribution $\mathbf{w} = \{w_i\}$ at time $n + 1$ is the same as the distribution at time n is called a *stationary distribution*. A Markov chain is called *irreducible* if it is possible to go from every state to every state in a finite number of steps, that is, there is always a path between any two states. A Markov chain is said to be aperiodic if it has no periodic state. A periodic state is a state that can be revisited by a path starting from it only at multiples of a given period. An irreducible and aperiodic Markov chain is called *ergodic*. For an ergodic Markov chain, the stationary distribution $\mathbf{w}$ exists and is unique, and $w_j = \lim_{n \to \infty} (\mathbf{P}^n)_{ij}$. The stationary distribution satisfies the left eigenvector equation $\mathbf{wP} = \mathbf{w}$. Thus, we can also think of the stationary distribution as a left eigenvector of the transition probability matrix $\mathbf{P}$ [30].

The entropy rate of a stationary Markov chain, with stationary distribution $\mathbf{w}$ and transition probability matrix $\mathbf{P}$, is given by

$$
\begin{aligned}
h^x &= \lim_{n \to \infty} \frac{1}{n} H(X_n | X_n - 1, \ldots, X_1) = \lim_{n \to \infty} H(X_n | X_{n-1}) \\
&= H(X_2 | X_1) = - \sum_{i=1}^{n} w_i \sum_{j=1}^{n} p_{j|i} \log p_{j|i}. \qquad (1.26)
\end{aligned}
$$

1.5   JENSEN–SHANNON DIVERGENCE

Many important inequalities and results in information theory are obtained from the concavity of the logarithmic function. A function $f(x)$ is convex over an interval $[a, b]$ (the graph of the function lies below any chord) if for every $x_1, x_2 \in [a, b]$ and $0 \le \lambda \le 1$,

$$f(\lambda x_1 + (1 - \lambda)x_2) \le \lambda f(x_1) + (1 - \lambda)f(x_2). \tag{1.27}$$

A function is strictly convex if equality holds only if $\lambda = 0$ or $\lambda = 1$. A function $f(x)$ is concave (the graph of the function lies above any chord) if $-f(x)$ is convex. For instance, x^2 and $x \log x$ (for $x > 0$) are strictly convex functions, and $\log x$ (for $x > 0$) is a strictly concave function.

Jensen's inequality can be expressed as follows. If f is a convex function on the interval $[a, b]$, then

$$\sum_{i=1}^{n} \lambda_i f(x_i) - f\left(\sum_{i=1}^{n} \lambda_i x_i\right) \ge 0, \tag{1.28}$$

where $0 \le \lambda \le 1$, $\sum_{i=1}^{n} \lambda_i = 1$, and $x_i \in [a, b]$. If f is a concave function, the inequality is reversed. A special case of this inequality is when $\lambda_i = 1/n$ because then

$$\frac{1}{n}\sum_{i=1}^{n} f(x_i) - f\left(\frac{1}{n}\sum_{i=1}^{n} x_i\right) \ge 0, \tag{1.29}$$

that is, the value of the function at the mean of the x_i is less than or equal to the mean of the values of the function at each x_i.

From this inequality, some properties of Shannon's information measures presented in Section 1.1 and Section 1.2 can be proved. The following properties [33] can also be derived:

1. $D_{KL}(p||q) \ge 0$ (Equation 1.8).

2. $D_{KL}(p||q)$ is convex in the pair (p, q).

3. $H(X)$ is a concave function of p.

4. Given X and Y with the joint distribution $p(x, y) = p(x)p(y|x)$, $I(X; Y)$ is a concave function of $p(x)$ for fixed $p(y|x)$ and a convex function of $p(y|x)$ for fixed $p(x)$.

The Jensen–Shannon divergence, derived from the concavity of entropy, is used to measure the dissimilarity between two probability distributions and has the important feature that a different weight can be assigned to each probability distribution. The *Jensen–Shannon divergence* (JS) is defined by

$$JS(\pi_1, \pi_2, \ldots, \pi_n; p_1, p_2, \ldots, p_n) = H\left(\sum_{i=1}^{n} \pi_i p_i\right) - \sum_{i=1}^{n} \pi_i H(p_i), \tag{1.30}$$

where $p_1, p_2, \ldots, p_n$ are a set of probability distributions defined over the same alphabet with prior probabilities or weights $\pi_1, \pi_2, \ldots, \pi_n$, fulfilling $\sum_{i=1}^{n} \pi_i = 1$, and $\sum_{i=1}^{n} \pi_i p_i$ is the probability distribution obtained from the weighted sum of the probability distributions $p_1, p_2, \ldots, p_n$. From the concavity of entropy, the Jensen–Shannon (JS) divergence [14] fulfills

$$JS(\pi_1, \pi_2, \ldots, \pi_n; p_1, p_2, \ldots, p_n) \geq 0. \tag{1.31}$$

The JS divergence measures how far the probabilities p_i are from their mixing distribution $\sum_{i=1}^{n} \pi_i p_i$, and equals zero if and only if all the p_i are equal. It is important to note that the JS divergence is identical to the mutual information $I(X; Y)$ when $\pi_i = p(x_i)$ (i.e., $\{\pi_i\}$ corresponds to the marginal distribution $p(X)$), $p_i = p(Y|x_i)$ for all $x_i \in \mathbb{X}$ (i.e., p_i corresponds to the conditional distribution of Y given x_i), and $n = |\mathbb{X}|$ [14, 111].

1.6 INFORMATION BOTTLENECK METHOD

The *information bottleneck method*, introduced by Tishby et al. [123], provides a quantitative notion of "meaningful" or "relevant" information and formulates a variational principle for the extraction of relevant information. More specifically, the objective of this technique is to extract a compact representation of the variable X, denoted by $\widehat{X}$, with minimal loss of mutual information with respect to another variable Y (i.e., $\widehat{X}$ preserves as much information as possible about the relevance or control variable Y). The variable Y must not be independent from the original variable X, that is, mutual information $I(X; Y)$ must be strictly positive.

Given an information channel between X and Y, the information bottleneck method tries to find an optimal trade-off between accuracy and compression of X when the bins of this variable are clustered. Soft [123] and hard [110] partitions of X can be adopted. In the first case, every $x \in \mathbb{X}$ can be assigned to a cluster $\hat{x} \in \widehat{\mathbb{X}}$ with some conditional probability $p(\hat{x}|x)$ (soft clustering). In the second case, every $x \in \mathbb{X}$ is assigned to only one cluster $\hat{x} \in \widehat{\mathbb{X}}$ (hard clustering).

In this book, we consider hard partitions and we focus our attention on the agglomerative information bottleneck method [110]. Given a cluster $\hat{x}$ defined by $\hat{x} = \{x_1, \ldots, x_l\}$, where $x_k \in \mathbb{X}$ for all $k \in \{1, \ldots, l\}$, and the probabilities $p(\hat{x})$ and $p(y|\hat{x})$ defined by

$$p(\hat{x}) = \sum_{k=1}^{l} p(x_k), \tag{1.32}$$

$$p(y|\hat{x}) = \frac{1}{p(\hat{x})} \sum_{k=1}^{l} p(x_k, y) \quad \forall y \in \mathbb{Y}, \tag{1.33}$$

the following properties are fulfilled:

1. The decrease in the mutual information $I(X;Y)$ due to the merging of $x_1, \ldots, x_l$ is given by

$$\delta I_{\hat{x}} = p(\hat{x}) JS(\pi_1, \ldots, \pi_l; p_1, \ldots, p_l) \geq 0, \qquad (1.34)$$

where the weights and probability distributions of the JS divergence (Equation 1.30) are given by $\pi_k = p(x_k)/p(\hat{x})$ and $p_k = p(Y|x_k)$ for all $k \in \{1, \ldots, l\}$, respectively. An optimal clustering algorithm should minimize $\delta I_{\hat{x}}$.

2. The information loss obtained by the merging of l components can be computed from the information loss obtained by $l-1$ consecutive mergings of pairs of components.

The information bottleneck method can be implemented by a greedy algorithm, where in each step the best possible merge is performed, that is, the merge that minimizes the loss of mutual information [110].

1.7 SUMMARY

This chapter presented the most important measures in information theory, which are entropy, conditional entropy, and mutual information. Entropy and mutual information, which express, respectively, the information content of a random variable and the correlation between two random variables, were analyzed in the context of an information (or communication) channel. It was shown that mutual information is a special case of the Kullback–Leibler distance, which is a measure of the divergence between two probability distributions. This chapter also introduced most of the basic concepts required for the development of the information-theoretic applications studied in this book: the mutual information decomposition, the entropy rate, the Jensen–Shannon divergence, and the information bottleneck method.

FURTHER READING

Butts, D.A. (2003). How much information is associated with a particular stimulus? *Network: Computation in Neural Systems,* 14:177–187.

Cover, T.M. and Thomas, J.A. (1991). *Elements of Information Theory.* Wiley Series in Telecommunications, John Wiley & Sons, New York.

Deweese, M.R. and Meister, M. (1999). How to measure the information gained from one symbol. *Network: Computation in Neural Systems,* 10(4):325–340.

Shannon, C.E. (1948). A mathematical theory of communication. *The Bell System Technical Journal,* 27:379–423, 623–656.

Slonim, N. and Tishby, N. (2000). Agglomerative information bottleneck. *Proceedings of NIPS-12 (Neural Information Processing Systems)*, MIT Press, 617–623.

Yeung, R.W. (2008). *Information Theory and Network Coding*. Information Technology: Transmission, Processing and Storage, Springer, New York.

CHAPTER 2

Visualization and Information Theory

CONTENTS

Visualization is a form of communication. Figure 2.1 shows a typical depiction of a general communication system considered by Shannon and Weaver as an abstract representation [106]. The source and destination of the message can be a person or a machine. The *encoder* (also referred to as transmitter) and *decoder* (also referred to as receiver) transform messages into signals and vice versa. Conceptually, the term "signal" is a generalization encapsulating messages represented by both digital and analog signals. In modern communication systems, we can simply consider both messages and signals as "data." Traditionally, the term "channel" refers to a transmission medium. In abstract, it is a function or process that operates on an input signal and sometimes adds noise, resulting in an output signal.

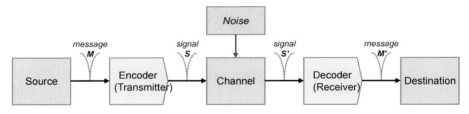

Figure 2.1 A general communication system [25].

It is not difficult to observe that a visualization process can have more or less the same abstraction. For instance, a person or a machine may encode some data using visual objects, resulting in a visualization to be transmitted via a display device. A viewer decodes the visualization, and makes sense of the data being conveyed.

Figure 2.2 depicts a general visualization pipeline, without human–computer, human–human, or inter-process interaction.[1] To make the comparison easier, we can group the eight processes, from *filtering* to *cognition* into three subsystems as shown in Figure 2.2, which we colloquially refer to as *vis-encoder*, *vis-channel*, and *vis-decoder*. This results in a model analogous to Figure 2.1.

---

[1]Interaction is an important part of visualization, and it can be accommodated in an information-theoretic framework for visualization. We will consider interactive visualization later in Sections 2.1.5, 2.2, 2.3.1, and 2.3.4.1.

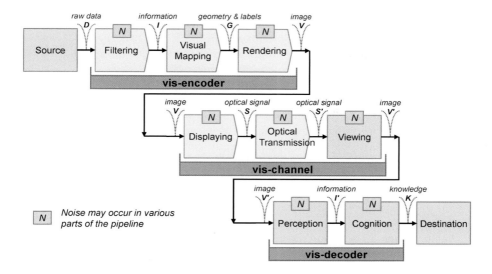

Figure 2.2 A general visualization system, which is decomposed into three subsystems, *vis-encoder*, *vis-channel*, and *vis-decoder* [25].

Information theory is "the science of quantification, coding and communication of information" [128]. It is generally agreed that information theory was founded by Shannon [106, 107] and Wiener [143]. We should also acknowledge those who suggested the basic ideas, including Nyquist [88], Hartley [60], and Fisher [46]. There is a strong connection between information theory and visualization. This chapter introduces readers to this connection.

2.1 INFORMATION-THEORETIC MEASURES IN VISUALIZATION

2.1.1 Alphabets and Letters

The term *data object* is an encompassing generalization of a family of terms that include datum, data point, data sample, data record, and dataset. It contains a finite collection of quantitative and/or qualitative measures that are values of a finite set of variables. For example, consider a univariate variable X for recording the population of a country. A value representing the UK population in 2010 is a datum, and thus a data object. A collection of the individual population figures of N countries in 2010 is also a data object, where the N values may be considered as a sample of data points of X, or separate records of N variables X_i ($i = 1, 2, \ldots, $N). Similarly, a time series that records the UK annual population between 1900 and 2010 is a data object. The 111 values in the time series may be considered as data points of the same univariate variable X, or a multivariate record for time-specific variables X_t ($t = 1900, 1901, \ldots, 2010$). Of course, the term *data object* can also refer to a multivariate data point that consists of values of conceptually different variables, e.g., area, population, and GDP of a country.

The generalization also encompasses datasets that are often regarded as "unstructured." For example, a piece of text may be treated as a multivariate record of M characters, each of which is a value of a variable C_j for encoding a letter, digit, or punctuation mark at a specific position j $(j = 1, 2, \ldots, M)$ within the text. Hence, the multivariate record is a data object. Alternatively, we can consider a composite variable, Y, which encodes all possible variations of texts with M or fewer characters. A specific text with $1 \leq k \leq M$ characters is thus a value of Y. This example also illustrates the equivalence between encoding a data object as a multivariate data record and encoding it as an instance of a single composite variable.

In this generalized context, let Z be a variable, and $\mathbb{Z} = \{z_1, z_2, \ldots, z_M\}$ be the set of all its valid values. Z may be a univariate, multivariate, or composite variable. When Z is a multivariate variable, each of its valid values, z_i, is a valid combination of valid values of individual univariate variables. When Z is a composite variable, we can flatten its hierarchy by encoding the hierarchical relationships explicitly using additional variables. The flattened structure thus represents a multivariate variable. Hereby z_i is a valid combination of valid values of individual variables including the additional ones. In information theory, such a set $\mathbb{Z}$ is referred to as an *alphabet*, and each of its member z_i as a *letter*.

When the probability of every letter, $p(z_i)$, is known or can be estimated, $p(Z)$ is the *probability mass function* for the set $\mathbb{Z}$. As detailed in Section 1.1, Shannon introduced the measure of *entropy*

$$H(Z) = -\sum_1^M p(z_i) \log p(z_i)$$

for describing the *level of uncertainty* of an alphabet. As the base of the logarithm is chosen to be 2, the unit of $H(Z)$ is the *bit*.

2.1.2 Quantifying Visual Information

Let us consider a simple, black-and-white time-series visualization as shown in Figure 2.3(left). Assume that the graph plotting area is given as 256×64 pixels. The time series has 64 independent samples, and each sample has an integer value range between 0 and 255. Samples are taken at a regular temporal step. The visualization displays 64 pixels corresponding to the random samples, and the connecting lines between consecutive samples. The probability mass function of each sampling value is independent and identically distributed, and we have $p = 1/2^8$. A visualization image may be associated with several different alphabets.

Data alphabet. Let S denote the random variable for a single sample, and Z denote a multivariate variable of 64 samples. $\mathbb{Z}$ is thus the alphabet encompassing all time series with 64 samples. Figure 2.3(left) shows an instance $z \in \mathbb{Z}$. The entropy for this variable Z is calculated as

$$H(Z) = \sum_{t=1}^{64} H(S_t) = -\sum_{t=1}^{64} \sum_{i=0}^{255} \frac{1}{2^8} \log \frac{1}{2^8} = 512. \tag{2.1}$$

This is very much expected as the time series requires a minimum of 64 bytes (i.e., 512 bits) to encode. In theory, we can also display this binary code as shown in Figure 2.3(right), which would require a huge perceptual and cognitive load to decode if not impossible. Figure 2.3(right) actually uses more than 1 pixel for each "bit" box, as the figure would otherwise be unreadable by human eyes. If we use only 4×4 pixels per "bit" box, it would result in a total of 2^{13} bits. In practice, we choose to use a graphical (or visual) mapping $G(Z)$ as in Figure 2.3(left), for which a plotting area with 256×64 black-and-white pixels (2^{14} bits) would be a minimal requirement. For this particular design of $G(Z)$, if it uses 2^{14} bits of display space, it is capable of depicting on average 512 bits of information.

If all of the samples take values only in the lower half of the value range, the entropy will be 448 instead of 512. Meanwhile, most users may sensibly halve the plotting area by remapping the y-axis from $[0-255]$ to $[0-127]$. The proportion of entropy reduction seems to differ from that of space reduction. It is interesting to know how well entropy relates to the visual display.

Graphics alphabet. The graphics alphabet consists of all possible time series that can use a visual representation, such as a line graph or a line of color-coded pixels. Different visual representations offer different capacities in conveying the data alphabet to be depicted. Assume that the data alphabet is sufficiently large for a static graphical mapping G. We define the average amount of information that G can depict as the *visualization capacity*, $V(G)$. G is usually constrained by a number of parameters, e.g., the required display space, the spatial partitioning of the display in relation to the data, use of colors, etc. Once these parameters are fixed, $V(G)$ is the entropy of a random variable that takes all possible distinguishable outputs of this specific mapping G.

For a specific input data variable X, we have

$$V(G) = \min\big(V(G(X)), H(X)\big),$$

where the "min" function indicates that a visualization *normally* cannot display more information than what is contained in the input data variable X. This follows the principle of *data processing inequality* in information theory [32]. However, in visualization, it is often advantageous to break the conditions of this principle. We will discuss this further in Section 2.3.1.

Display space capacity. We define the display bandwidth available for visualization as the *display space capacity*, D, which takes into account the number of pixels in the display, and the depth of each pixel. Note that D is

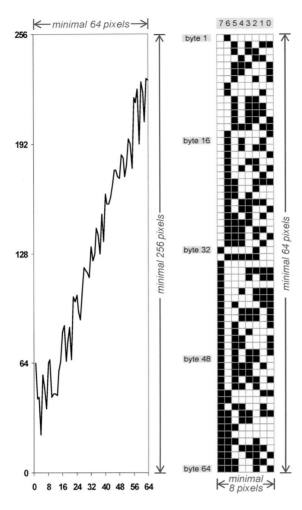

Left: A 64-sample time series with possible integer values ranging between 0 and 255 is plotted as a line graph. A display space with a minimum of 256 × 64 pixels will be necessary to depict all data instances (i.e., letters) possibly contained in the data alphabet.

Right: The same time series can in theory be displayed using 64 × 8 pixels. In practice, more pixels are needed to make individual squares more distinguishable.

Figure 2.3 Two different visual representations of a time series [25].

independent of a data variable X or a graphical mapping G. It indicates the maximum entropy achievable by any graphical mapping within this display space.

Here we do not assume that $H(X)$ and $V(G)$ are defined upon a uniform probability distribution, as a priori knowledge about the distribution is usually an essence of a design process in visualization. Later in Section 2.1.4, we will show an example of how a non-uniform probability mass function leads to a commonly used visual representation.

Three relative measures: VMR, ILR, and DSU. The quantities V and

D have the same unit as entropy H. This allows us to define the following measurements:

$$\text{Visual mapping ratio (VMR)} = \frac{V(G)}{H(X)}. \tag{2.2}$$

$$\text{Information loss ratio (ILR)} = \frac{\max\big(H(X) - V(G), 0\big)}{H(X)}. \tag{2.3}$$

$$\text{Display space utilization (DSU)} = \frac{V(G)}{D}. \tag{2.4}$$

Here we assume $H(X) > 0$. When $H(X) = 0$, there is no uncertainty or information in X [32]. Similarly, we assume $D > 0$.

Using the time-series visualization in Figure 2.3(a) as an example, we have $H(Z) = 512 = 2^9$, $V(G) = 2^9$, and $D = 2^{14}$. Hence, VMR $= 1$, ILR $= 0$, and DSU $= 2^{-5} = 0.03125$.

For the above example of remapping the y-axis from $[0 - 255]$ to $[0 - 127]$, we can obtain $H(Z) = 448 = 7 \times 2^6$, $V(G) = 7 \times 2^6$, and $D = 2^{13}$. Hence, the entropy and visualization capacity proportionally reduce the same number of bits. Note that VMR, ILR, and DSU are measurements of ratios, and thus unitless.

From the perspective of data compression, coding an 8-bit value using 2^8 bits for each sample seems totally insane. Nevertheless, it makes sense for visualization. The human visual system is a highly parallel system. It takes a single viewing step to determine that in Figure 2.3(a) has a starting value of 64. It would take 8 viewing steps, together with much cognitive reasoning, to establish this fact from Figure 2.3(b). There is thus no reason to be apprehensive about the apparent "inefficiency" of visualization from the perspective of data compression. This suggests that application of information theory to visualization needs to accommodate and address a different emphasis. We will continue this discussion of how the apparent inefficiency is used in visual multiplexing in Section 2.4.1 and such inefficiency is featured in cost–benefit analysis in Section 2.3.4.2.

Potential information loss. However, often we do not have 2^k bits of display space for every k-bit value, as we are usually constrained by a limited number of pixels available in most practical applications. In such a situation, the above measurements provide us with a quantitative evaluation of a visual design G. For instance, let us reduce the display space from 64×256 pixels to 64×64 pixels. We denote the new DSU as D', which is 2^{12} bits. The geometry mapping of the original visual design G has to be modified. The new design G' may not be able to depict the full amount of raw data.

Consider a simple data mapping function, M, that maps the above-mentioned time-series data variable Z to a new data variable Z', where each value $j \in Z$ is mapped to $i \in Z'$ such that $i = \lfloor j/4 \rfloor$. We have

$$V(G') = H(Z') = H(M(Z)) = -\sum_{1}^{64}\sum_{0}^{63} \frac{1}{2^6} \log \frac{1}{2^6} = 384. \qquad (2.5)$$

Hence, $\text{VMR}' = 384/512 = 0.75$, $\text{ILR}' = (512 - 384)/512 = 0.25$, and $\text{DSU}' = 384/2^{12} \approx 0.094$. Figure 2.4(a) shows a visualization of $M(z)$, where $z \in Z$ is the same time series as in Figure 2.3(a). In Section 2.1.4, we will discuss the relative merits of several visual mappings for displaying data variables that have different probability mass functions on the same 64×64 display.

2.1.3 Information Sources and Communication Channels

In information theory and its application of data communication, the terms *information sources* and *communication channels* are formally defined and categorized. We can adopt or adapt these concepts from the perspective of visualization.

2.1.3.1 Information Sources

An *information source* is a process that produces a message or a sequence of messages to be communicated. A source is said to be "memoryless" if each message is an instance of an independent random variable and stochastically obeys the same probability distribution as the other messages. A source is said to be *stationary* if its probability distribution is spatially and temporally invariant. In other words, when the source is moved in space or its message generation is shifted in time, its probability distribution does not change. In information theory, these two properties are the preconditions of many theorems [56]. The preconditions are also commonly assumed by most communication systems and compression algorithms.

A visualization system may encounter three types of information sources, namely *input data*, *interaction*, and *pre-stored knowledge*. If we focus only on the input data, the parallel between visualization and communication is apparent (cf. Figures 2.1 and 2.2). In most cases, assuming that the input is a memoryless and stationary source is a sensible abstraction for both theoretic and algorithmic development.

Interaction allows users to provide a visualization system with additional information, such as viewing parameters and mapping functions, resulting in different output data.

Pre-stored knowledge includes hard-coded knowledge (e.g., feature recognition) in the vis-encoder, and human knowledge in the vis-decoder (Figure 2.2). The former normally is not of a stochastic nature, but this may change in future systems with the introduction of knowledge-assisted visualization [22]. The latter is stochastic, especially when considering the whole population of potential users of a system.

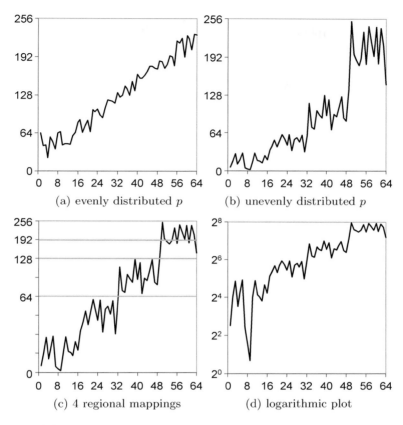

(a) evenly distributed p

(b) unevenly distributed p

(c) 4 regional mappings

(d) logarithmic plot

Figure 2.4   (a) Reduced display may result in information loss. When displaying the same time series in Figure 2.3(a) using only 64x64 pixels, the *visualization capacity* V is reduced from 512 to 384 bits, resulting in 25% information loss. (b) For a data alphabet with a non-uniform probability mass function p, the *visualization capacity* V is further reduced to 368, while the entropy of the input data is now 496. (c) Using 4 regional spatial mappings, one can improve V and reduce the information loss. However, though this is entropy based, the transitions between the 4 data ranges are not continuous. (d) One can replace (c) with a logarithmic plot to maintain the continuous spatial transition [25].

There is still a major scientific gap in understanding the probabilistic properties of human interaction and human knowledge. Many existing theorems in information theory may not be readily applicable when such information sources are considered, and some adaptation and extension of information theory will be necessary. This does not in any way suggest that information theory should not become a theoretic framework of visualization. More appropriately, it suggests that visualization offers bountiful phenomena that can stimulate the advancement of information theory.

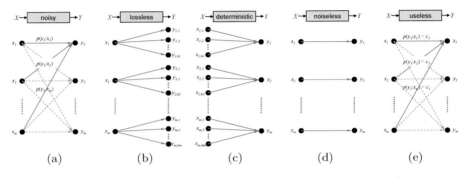

Figure 2.5 Five common types of communication channels [25].

2.1.3.2 Channels

A *channel* is the medium over which coded messages are transmitted from the encoder to the decoder. Here, we consider only discrete channels. In communication, as illustrated in Figure 2.1, unintended changes could be made to a coded message, resulting in errors in transmission. Such changes are referred to as *noise* or *perturbation*. A channel may have properties such as bandwidth, transmitted power, and error rate.

Let X and Y denote the random variables corresponding respectively to the input and output of a channel. Figure 2.5(a) shows a typical noisy channel, where an input message x_i may be received as y_j with a probability mass function $p(y_j|x_i)$. As defined in Section 2.1.1, in abstract, different messages of a variable are different letters in the corresponding alphabet. For a noisy channel, there is a many-to-many mapping between the letters in the input alphabet $\mathbb{X}$ and those in the output alphabet $\mathbb{Y}$.

In a visualization system (Figure 2.2), each subsystem or processing component has its own input and output alphabets $\mathbb{X}$ and $\mathbb{Y}$, where letters x_i and y_i represent possible instances of data objects. For example, considering a vis-encoder subsystem for the time-series visualization in Figure 2.3(a), we have $\mathbb{X} = \mathbb{Z}$ for the alphabet encompassing all possible combinations of raw data and $\mathbb{Y}$ is the alphabet encompassing all possible results of graphics mappings $Y = G(Z)$. When we study a particular algorithmic component of a subsystem, e.g., a filtering or clustering function, X and Y can also represent the input and output variables related only to this component.

Many processes in a visualization pipeline are noisy channels. In a vis-decoder subsystem, a graphical object depicted in a visualization can easily result in different interpretations by different viewers. Using the notation in Figure 2.5(a), an instance x_1, for example, may be probabilistically interpreted as different instances $y_i \in Y$, with different $p(y_i|x_1), i = 1, 2, \ldots, m$. When x_1 is intended to be seen as a specific y_k, we would like to maximize $p(y_k|x_1)$. This can be achieved by a better design of the visual mapping, or by helping the viewers to detect and correct errors in the vis-decoder subsystem.

A channel is said to be *lossless* if every input letter can be uniquely determined from an output letter as illustrated in Figure 2.5(b). Though it is a one-to-many mapping for each x_i, the mapped letters $\{y_{i,1}, y_{i,2}, \ldots, y_{i,k_i}\}$ are grouped into a set corresponding to x_i uniquely. Such a channel has a conditional entropy $H(X|Y) = 0$ for all input distribution. When a lossless channel introduces *redundancy*, it provides a mean for error detection and correction in communication. We will examine this in Section 2.1.5.

A channel is said to be *deterministic* if every input letter uniquely determines an output letter, as shown in Figure 2.5(c). Such a channel has a conditional entropy $H(Y|X) = 0$ for all input distribution, and can facilitate abstraction. For example, quantization at different stages of the pipeline, such as color mapping, behaves as a deterministic channel. Meanwhile, perceptually retrieving values from colors, i.e., the reverse mapping, is not deterministic.

A channel is said to be *noiseless* if it is lossless and deterministic, resulting in a one-to-one mapping between input and output letter as illustrated in Figure 2.5(d). In visualization, such a channel is only desirable when the input data alphabet is small. For large scale data visualization, a totally noiseless visualization system may be neither practical nor helpful.

In visualization, an abstraction process often does not throw the original data away, and the abstraction may be used merely to support visual mapping, e.g., assigning colors to data points in different clusters. In other words, such a process is a combination of deterministic and noiseless channels. This mechanism is commonly used in supporting visual categorization, and focus of attention.

A channel is said to be *useless* if every output letter has an equal chance of resulting from any input letter, as shown in Figure 2.5(e). In terms of entropy, we have $H(X|Y) = H(X)$.

A discrete channel is said to be *memoryless* if the probability distribution of the output depends only on the input at that time, and is independent of previous inputs and outputs of the channel [32]. The channels in any vis-decoder are certainly not memoryless, and interaction introduces historical dependency. This hinders the direct application of some major theorems in information theory, e.g., Shannon's channel coding theorem [107]. Nevertheless, the basic idea for handling errors in noisy channels is very much applicable to visualization.

2.1.4 Visual Coding in Noiseless Channels

In data communication, coding schemes are broadly divided into two main categories, namely *source coders* and *channel coders*. A source coder focuses on the messages from the source, and tries to find the most compact representation of the messages. A channel coder focuses on the noise on the channel, and tries to find a cost-effective representation that can help detect or/and correct errors introduced by the channel. We study these two categories in the context of visualization in this section and Section 2.1.5 respectively.

Many encoding schemes in data communication can inspire us to develop new data abstraction and visual encoding techniques. In this section we give one example of such schemes to illustrate the relevance of source coding to visualization. We call this scheme *entropy-based spatial mapping*, which conceptually bears a strong resemblance to *entropy coding* in data compression and communication. The latter is a family of coding schemes, where fixed-length codewords are replaced with variable-length codewords. The well-known entropy encoding schemes include Huffman encoding, Shannon–Weaver–Fano encoding, and arithmetic encoding.

Recall the time-series example in Figure 2.4(a) that was discussed in Section 2.1.2. The original data variable is Z, and its corresponding alphabet is $\mathbb{Z} = \{z_{i,j} | i = 1, 2, \ldots, 64; j = 0, 1, \ldots, 255\}$. We now consider a data alphabet $\mathbb{W}$ that has a different probability mass function. Its value range $[0, 255]$ is divided into 2^d equal-sized sub-ranges, $R_1, R_2, \ldots, R_{2^d}$, where d may take an integer value between 0 and 8. Each sub-range thus has 2^{8-d} possible data values. Figure 2.4(a) is an instance when $d = 0$, after a linear mapping of the value range from $[0, 255]$ to $[0, 63]$. When $d > 0$, the probability mass function, $p(w_{i,j})$, varies according to the sub-ranges. Suppose that there is a $1/2^k$ chance that the sample values will fall into sub-range $R_k, k = 1, \ldots, 2^d - 1$. The chance for sub-range R_{2^d} is the remainder of probability, i.e., R_{2^d} and R_{2^d-1} have the same probability.

For example, when $d = 2$, we have four sub-ranges, with probabilities, $1/2$, $1/4$, $1/8$, and $1/8$ respectively. Figure 2.4(b) shows the visualization of such a time series, which uses the same linear mapping from $[0, 255]$ to $[0, 63]$ as in Figure 2.4(a).

Assume that $p(w_{i,j})$ within each sub-range is independent and identically distributed. We thus have $p(w_{i,j}) = 1/2^7$ in R_1, $1/2^8$ in R_2, and $1/2^9$ in R_3 and R_4. The entropy of the corresponding variable, $H(W)$, is the sum of entropies of the four sub-ranges. The *visualization capacity* for the linearly mapped data, $V(G_l(W))$, is the sum of the visualization capacities of those sub-ranges. We have $H(W) = 496$ and $V(G_l(W)) = 368$.

Hence, we have visual mapping ratio $\text{VMR}_l = 368/496 \approx 0.742$, information loss ratio $\text{ILR}_l = (496 - 368)/496 \approx 0.258$, and display space utilization $\text{DSU}_l = 368/2^{12} \approx 0.09$. In comparison with a uniform distribution (last three paragraphs of Section 2.1.2), the *visualization capacity V* is slightly reduced

Table 2.1 Quantities and measures of different numbers of sub-ranges. When the number is $2^d, d > 1$, the probability mass function varies logarithmically in different sub-ranges.

# Sub-ranges	1, 2	4	8	16	32
entropy H	512	496	447	384	320
linear $V(G_l)$	384	368	319	256	192
linear VRM$_l$	0.750	0.742	0.714	0.667	0.600
linear ILR$_l$	0.250	0.258	0.286	0.333	0.400
linear DSU$_l$	0.094	0.090	0.078	0.062	0.047
non-linear $V(G_{nl})$	—	384	384	384	384
non-linear VRM$_{nl}$	—	0.774	0.859	1.000	1.200
non-linear ILR$_{nl}$	—	0.226	0.141	0.000	0.000
non-linear DSU$_{nl}$	—	0.094	0.094	0.094	0.094

(cf. 0.75), while there is slightly more information loss (cf. 0.25), and poorer utilization of display space (cf. 0.094).

Let us consider a non-linear mapping function that maps $R_1 : [0, 63]$ to $[0, 31]$, $R_2 : [64, 127]$ to $[32, 47]$, $R_3 : [128, 191]$ to $[48, 55]$, and $R_4 : [192, 255]$ to $[56, 63]$. It is not difficult to derive that $V(G_{nl}(W)) = 384$ with VMR$_{nl} \approx$ 0.774, ILR$_{nl} \approx 0.226$, and DSU$_{nl} \approx 0.094$. We have slightly improved the *visualization capacity*, and as well reduced information loss. When we plot out G_{nl} in Figure 2.4(c), this is conceptually similar to a logarithmic mapping in Figure 2.4(d).

If we increase the number of sub-ranges, for instance, for $d = 3, 4, 5$, the improvement becomes more significant and interesting as shown in Table 2.1. For $d = 3, 4, 5$, the six lower data ranges are mapped to six visualization ranges with $32, 16, 8, 4, 2, 1$ pixels respectively. The remaining high-value data ranges are mapped to a single 1-pixel range. The non-linear mapping manages to maintain the *visualization capacity* at 384. The higher d is, the closer the distribution is to a logarithmic distribution. In other words, logarithmic plots are, in effect, a means to increase *visualization capacity* V when the distribution of the sample values follows a certain logarithmic pattern. This explains why logarithmic plots are commonly used in the sciences and engineering.

2.1.5 Visual Coding in Noisy Channels

In visualization, *interaction* is a primary means for helping a viewer detect and correct errors. For example, in medical visualization, a volumetric model (e.g., a CT scan dataset) is often displayed using direct volume rendering. In a resulting visualization, as illustrated in Figure 2.6, samples at different depth along the same ray are projected onto the same pixel, and the colors and opacities of these samples are combined according to the volume rendering integral, resulting in a pixel color that captures information from many samples.

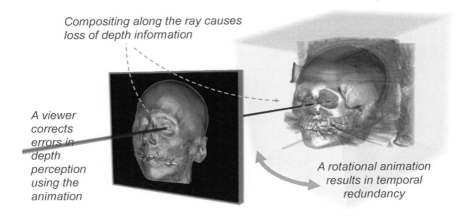

Compositing along the ray causes
loss of depth information

A viewer
corrects
errors in
depth
perception
using the
animation

A rotational animation
results in temporal
redundancy

Figure 2.6 Motion parallax, facilitated by interaction and real-time rendering, is a form of error correction in visualization [25].

The process itself is very similar to frequency-division multiplexing in communication, where different signals are transmitted in several distinct frequency ranges over that same medium. However, in the case of volume visualization, we cannot assure the distinctive separation between different frequency ranges, though a good transfer function may provide more visual cues to alleviate the problem. Furthermore, there is a substantial loss of 3D information in a 2D projection. In other words, viewers are expected to make mistakes in determining the shapes depicted in such a visualization.

Typically, a viewer interactively rotates the volume, receiving multiple visualizations for the same model. From these interactively generated visualizations, the imprecise models perceived initially gradually converge to a more accurate 3D model in the viewer's mind. Conceptually, this is similar to *backward error correction* (or *automatic repeat request*) in communication, which requests retransmission of erroneous data. The changes of viewing positions spread errors across different parts of the model, making error detection and correction possible for each individual part. This is conceptually very similar to multidimensional parity-check coding, which is a block-based error correction scheme.

Multiview visualization is another means for error detection and correction, especially in visualizing non-spatial data, where errors are often due to the perceptual load of visual search, change detection, and attention. Multiview visualization allows the same information to be found in different views, increasing the probability of locating the information. This is very similar to repetition coding schemes in data communication.

This naturally leads to the issue of *redundancy*. All error detection or correction coding schemes cause redundancy. Rheingans and Landreth studied the benefits of redundancy in visualization [95]. They found that (i) different parameters of a visual mapping convey different types of information with different efficiency, (ii) multiple display parameters can overcome visual deficiencies; and (iii) multiple display parameters reinforce each other. Information theory can provide support to their conclusions.

Considering that the *vis-decoder* part of the pipeline in Figure 2.2 is highly noisy, it is a great challenge to design visual mappings with built-in error detection and correction mechanisms.

2.2 INFORMATION-THEORETIC LAWS FOR VISUALIZATION

One major advantage of underpinning visualization using the information-theoretic framework is the potential of applying or adapting laws (and theorems) for articulating guiding principles in visualization. In this section, we consider three basic laws in information theory, and translate them to rules in visualization. Later in Section 2.3, we will show the mathematical connection between a benefit measurement for visualization and a theorem in information theory.

In visualization, it is common that many events are inter-related. For example, in interactive exploration, a user may first obtain an overview visualization $G_{overview}$ of a dataset, and then apply a zoom-in operation, resulting in one of the possible detailed views $G_{detail[i]}$. The information contained in $G_{detail[i]}$ is thus related to that in $G_{overview}$.

Let X and Y be two random variables with a joint probability mass function $p(x, y)$. $H(X, Y)$ denotes the *joint entropy* of the two variables; $H(Y|X = x)$ denotes the *conditional entropy* of Y given that X is known to be x; and $H(Y|X)$ denotes the *conditional entropy* of Y for all possible events in X (see Section 1.1).

Consider that X and Y model the probabilistic attributes of $G_{overview}$ and $G_{detail[i]}$ respectively. Here X and Y represent two variables associated with the two visualization alphabets, $G_{overview}(Z)$ and $G_{detail[i]}(Z)$, where Z is the input data variable shared by both views. We can apply some fundamental laws in information theory to explain different phenomena in the overview-detail situation. Below are examples of applying two such laws to articulate rules for zooming interaction.

Rule 1. $H(X, Y) \leq H(X) + H(Y)$ [2]. The joint uncertainty of the two views does not exceed the sum of the uncertainty exhibited by each individual view. In other words, having two views can reduce uncertainty. The equality is valid only when X and Y are independent, i.e., $G_{overview}$ and $G_{detail[i]}$ are not related to each other.

Rule 2. $H(Y|X) \leq H(Y)$ [56]. In the overview-detail situation, the possible variations of $G_{detail[i]}$ is strongly governed by those of $G_{overview}$. As illustrated in Figure 2.7, the distribution and orientation of the visual primitives in the overview determine the overall trend of the distribution and orientation of those in the detailed view. If the event of the square box to be zoomed is known, the entropy of $G_{detail[i]}$ is reduced significantly. After having seen the overview, the viewer has a rough idea about the detailed view $G_{detail[i]}$, and hence is less uncertain about it. Information-theoretically, this means $H(Y|X = x) < H(Y)$.

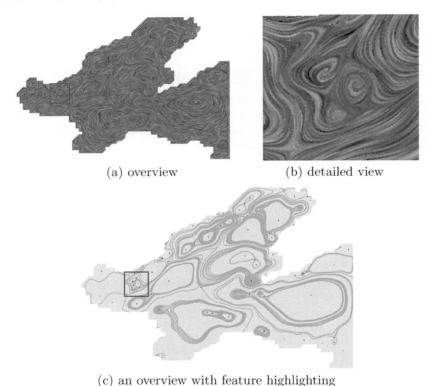

$\quad$ (a) overview $\qquad\qquad\qquad\qquad$ (b) detailed view

(c) an overview with feature highlighting

Figure 2.7 An example of mutual information in flow visualization. The overview in (a) may not show enough information to encourage a user to explore the detailed view in (b). Using feature extraction and highlighting, the new overview in (c) contains more feature-related mutual information about (b) [25].

$\quad$ The notion of conditional entropy exists inherently in human-centered processes, that is, the *vis-decoder* subsystem in Figure 2.2. In particular, this rule underpins the design principle for interactive exploration with overview, zoom, and detailed views [109].

$\quad$ In situations where $H(Y|X = x)$ is not as low as the viewer expected, the viewer would be either confused or surprised. For example, if a visualization system did not follow the basic design guideline that a zoom operation should ensure a succeeding view G_{k+1} has the same orientation of the preceding view G_k, the uncertainty about G_{k+1} will be significantly increased. In the case of Figure 2.7, most viewers would be confused when encountering a rotated Figure 2.7(b) (e.g., by 90°).

Alternatively, a viewer may be surprised to see some vortices in Figure 2.7(b) as there may not be a hint of its existence in Figure 2.7(a). In such a situation, the instinctive expectation of a higher conditional entropy is disadvantageous, especially in very large dataset visualization. Much research effort has been made to extract important features and highlight such features in higher-level views (e.g., [64]).

Mutual information $I(X;Y)$ measures the amount of the reduction of uncertainty of one random variable X due to the knowledge of another Y (see Section 1.2). In the case of overview first and details on demand, the reduction is through the visualization $G_{overview}$, which contains information about $G_{detail[i]}$. In information theory, there are a number of fundamental rules about mutual information, which are applicable to visualization events. The next rule is such an example.

Rule 3. $I(X;Y) = I(Y;X)$ [32]. This implies that the information about $G_{overview}$ in $G_{detail[i]}$ is the same as that about $G_{detail[i]}$ in $G_{overview}$. Undoubtedly, in most cases, the information about $G_{detail[i]}$ resides primarily in the corresponding window in $G_{overview}$. Let us partition $G_{overview}$ into n disjoint windows, each corresponding to a $G_{detail[k]}, k = 1, 2, \ldots, n$. We have:

$$\sum_{k=1}^{n} I(G_{detail[k]}; G_{overview}) = \sum_{k=1}^{n} I(G_{overview}; G_{detail[k]}). \qquad (2.6)$$

The left-hand side represents the total mutual information about all detailed views in an overview, while the right-hand side represents the total mutual information about the overview in all n detailed views. The former corresponds to one viewing step, but the latter corresponds to n viewing steps. This confirms the principle of overview first and details after [109].

Mutual information can also be used to quantify the effectiveness of a type of visualization. Consider a simplified example, where a viewer makes a decision to zoom-in on the box in Figure 2.7(a) based on whether there is a hint of vortices in the box. Let A be a random variable with two possible states, "show hints" and "show no hint" in that box. Similarly, let B be a variable about the detailed view of the box, with two possible states, "show at least one vortex" and "show no vortex."

Table 2.2 shows an example joint probability mass function $p(a, b)$ in columns 3 and 4. We obtain $I(B; A) \approx 0.147$, indicating the amount of uncertainty of Figure 2.7(b) that can possibly be reduced by having visualized Figure 2.7(a). Suppose that we introduce a feature-highlighting technique to improve Figure 2.7(a), as shown in Figure 2.7(c). C is a random variable similar to A but corresponds to Figure 2.7(c). The new probability mass function $p(c, b)$ is given in columns 6 and 7 of Table 2.2. The technique results in a 40% probability (instead of 25% previously) of showing a hint of any vortex in the

Table 2.2 This is an illustrative example in relation to Figure 2.7. Two joint probability mass functions $p(a,b)$ and $p(c,b)$ represent the likelihood that showing a hint of a vortex in either (a) or (c) may lead to showing an actual vortex in (b).

	$p(b)$ ▼	$p(a,b)$		$p(b)$ ▼	$p(c,b)$	
		hint	no hint		hint	no hint
vortex	0.5	0.25	0.25	0.5	0.4	0.1
no vortex	0.5	0.05	0.45	0.5	0.1	0.4
	$p(a)$ ▶	0.3	0.7	$p(c)$ ▶	0.5	0.5

same box area. Although the false positive has increased from 5% to 10%, the mutual information $I(B;C)$ has risen to about 0.278. In other words, Figure 2.7(c) can now tell more about Figure 2.7(b) in terms of vortices.

2.3 INFORMATION-THEORETIC PROCESS OPTIMIZATION

In the literature, many visualization pipelines have been proposed (e.g., [127, 129, 71]). Figure 2.2 is one such example. In practice, a visualization workflow normally includes machine-centric components (e.g., statistical analysis, rule-based or policy-based models, and supervised or unsupervised models) as well as human-centric components (e.g., visualization, human–computer interaction, and human–human communication). The integration of these two types of components becomes more and more common since *visual analytics* [122, 144] has become a de facto standard approach for handling large volumes of complex data.

   In the abstract, we may thus consider a visualization workflow as an instance of a general data processing workflow as depicted in Figure 2.8. The workflow, which is decomposed into L processing steps, transforms "data" into some "decisions" made by humans or machines. Here the term *decision* is used to encompass many forms of deliberation, such as observation, judgment, opinion, hypothesis, and so on. The deliberations may or may not yield new conclusions, and may or may not lead to actions. Hence, "no new conclusion" is also regarded as a decision, and "no action" is also regarded as an action. Although the workflow in Figure 2.8 is a sequential system, most everyday parallel workflows and iterative/interactive workflows can be sequentialized. This relates to the *sequentialization theorem* in theoretical computer science [48].

   Given a visualization workflow in a specific context, it is inevitable that one would like to improve its cost–benefit ratio, from time to time, in relation to many factors such as accuracy, speed, computational and human resources, credibility, logistics, changes in the environment, data or tasks concerned, and so forth. Such improvement can typically be made through introducing new

Figure 2.8 A general data processing workflow.

technologies, restructuring the existing workflow, or re-balancing the trade-off between different factors. While it is absolutely essential to optimize each visualization workflow in a heuristic and case-by-case manner [87], it is also desirable to study the process optimization theoretically and mathematically through abstract reasoning. In many ways, this is similar to the process optimization in tele- and data communication, where each subsystem is optimized through careful design and customization but the gain in the cost–benefit ratio is mostly underpinned by information theory [32, 107]. In this section, we examine, in abstraction, the process optimization in visualization from an information-theoretic perspective.

Visualization is a form of information processing. Like other forms of information processing (e.g., statistical inferences), visualization enables transformation of information from one representation to another. The objective of such a transformation is typically to infer a finding, judgment, or decision from the observed *data*, which may be incomplete and noisy. The input to the transformation may also include "soft" *information* and *knowledge*, such as known theories, intuition, belief, value judgment, and so on. Another form of input, which is often referred to as *priors*, may come from knowledge about the system where the data are captured, facts about the system or related systems, previous observations, experimentations, analytical conclusions, etc. Here we use the terms *data*, *information*, and *knowledge* according to the commonly used definitions in computational spaces [22].

All inferential processes are designed for processing a finite amount of information. In practice, they all encounter some difficulties, such as the lack of adequate technique for extracting meaningful information from a vast amount of data; incomplete, incorrect, or noisy data; biases encoded in computer algorithms or biases of human analysts; lack of computational resources or human resources; urgency in making a decision; and so on. All inferential problems are inherently underdetermined problems [50, 51].

The traditional machine-centric solutions to the inferential problem address these difficulties by imposing certain assumptions and structures on the model of the system where the data are captured. If these assumptions were correctly specified and these structures were perfectly observed, computed inference based on certain statistics (e.g., moments) would provide us with perfect answers. In practice, it is seldom possible to transform our theory, axioms, intuition, and other soft information into such statistics. Hence, optimization of a visualization process is not just about the best statistical method, the best analytical algorithm, or the best machine-learning technique. It is also about the best human-centric mechanisms for enabling uses of "soft" information and knowledge.

2.3.1 Data Processing Inequality

As discussed in Section 2.1.2, a visual representation of a dataset may take more space than the dataset itself. However, in principle, the visual representation does not include more information than what is in the original dataset. In general, any step in a data processing pipeline is considered unable to generate more information than what is in the original data variable. In information theory, this is referred to as *data processing inequality* [32], which states: if random variables X, Y, Z form a Markov chain in the order of $X \rightarrow Y \rightarrow Z$, then we have the following inequality between their mutual information:

$$I(X;Y) \geq I(Y;Z). \tag{2.7}$$

X, Y, Z are said to form a Markov chain if Z depends only on Y and is conditionally independent of X.

If we only allow data input as the information source for the channels in the chain, this inequality stands. However, this principle should not be naively applied to all visualization workflows, because most of such workflows are not Markov chains.

First, many processes in visualization are interactive processes, so we cannot guarantee that Z solely depends on Y. Second, even if there is no interaction, we usually take into account our knowledge about the raw data (e.g., X), when we design an algorithm at a late stage of the chain (e.g., for $Y \rightarrow Z$). Z is not conditionally independent of X. The condition for a Markov chain is thus broken.

In fact, we should not be disappointed by the fact that the data processing inequality is not as ubiquitous in visualization as one would expect. This fact only implies that interaction and domain knowledge about the raw data are critical in breaking the bottleneck of "data processing inequality." This explains why most visualization systems are interactive systems, and supports the argument for knowledge-assisted visualization [22]. The example in the following section illustrates the apparently rapid information loss in a data processing workflow if it were considered to be a Markov chain.

2.3.2 Transformation of Alphabets in a Visualization Process

In many data-intensive environments, the alphabet of raw input data may contain numerous letters. For example, consider all valid time series of share prices within a one-hour period. Assuming that the share price is updated every 5 seconds, there are 720 data points per time series within an hour. Assuming that we represent share price at USD $0.01 resolution using 32-

bit unsigned integers,[2] the minimum and maximum values are 0 and $2^{32} - 1$ cents respectively. If the probability of different time series were uniformly distributed, the entropy of this alphabet would be $23040 = 720 \times \log(2^{32})$ bits. This is the *maximal entropy* of this alphabet. In practice, as many high values in the range $[0, 2^{32} - 1]$ are very unlikely, and sudden changes between a very low value and a very high value (or vice versa) during a short period are also rare, the actual entropy is lower than 23040 bits.

On the other hand, if we need to consider r of such time series in order to make a decision, the size of the new alphabet will increase significantly. Although some combinations of r time series may be highly improbable, they may still be valid letters. Hence the maximal entropy of this new alphabet is $23040r$ bits. Let us consider such r time series as the initial raw data for a data analysis and visualization process as illustrated in Figure 2.9.

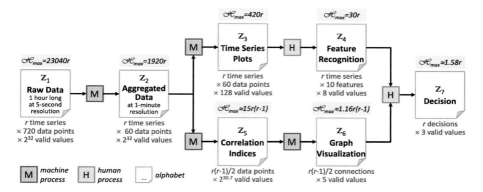

Figure 2.9 An example transformation of alphabets during a data analysis and visualization process. From left to right, the initial alphabet corresponds to r time series, each capturing a share price at 5-second intervals within an hour. For each time series, the 12 data points in every minute are then aggregated into a mean value. The r time series is then visualized as line plots. The analyst identifies various features during the visualization, such as different levels of rise or fall, different intensity, etc. Meanwhile, the analyst computes the correlation indices between each pair of time series and visualizes these using, for instance, a circular graph plot, where correlation indices are mapped to five different colors. The analyst finally makes a decision for each of the r shares as to buy, sell, or hold. The maximal entropy H_{max} shows a decreasing trend from left to right [24].

---

[2]By the end of 2014, the highest share price in the US is probably that of Berkshire Hathaway Inc. (BRK-A) at 22,937,400 cents, i.e., $2^{24} < 22,937,400 < 2^{25}$). Source: Yahoo Finance.

One may find that the resolution of 1 data point per 5 seconds is not necessary, and choose to reduce it to 1 data point every minute by computing the average of 12 data points in each minute. The average values may also be stored using 32-bit unsigned integers. This aggregation results in a new alphabet, whose maximal entropy of $1920r = r \times 60 \times \log(2^{32})$ bits. As indicated in Figure 2.9, one may use line plots with 128 distinguishable pixels along the y-axis. When we use these plots to visualize these r time series, we may only be able to differentiate up to 128 data values per data point. In this case, the maximal entropy is reduced to $r \times 60 \times \log(128) = 420r$ bits.

When one observes these r time series, one may identify some specific features, such as [rise, fall, or flat], [slow, medium, or fast], [stable, uneven, or volatile] and so on. These features become a new set of variables defined at the level of an hour-long time series. If we construct a new alphabet based on these feature variables, its entropy will be much less than $23040r$ bits. For example, if there are 10 feature variables, each with 8 valid values, the maximal entropy of this "observational" alphabet is $30r$ bits.

When one analyzes the relations among these r time series, one may, for instance, compute the correlation indices between every pair of time series. This yields $r(r-1)/2$ numbers. Assuming that these are represented using 32-bit floating-point numbers, the maximal entropy of this "analytical" alphabet is around $15r(r-1)$ bits as the single precision floating-point format supports some $2^{30.7}$ values in $[-1, 1]$. When we visualize these correlation indices by mapping them to, for instance, five colors representing $[-1, -0.5, 0, 0.5, 1]$, the entropy is reduced to $\log(5)r(r-1)/2 \approx 1.16r(r-1)$ bits.

One may wish to make a decision with three options, [buy, sell, or hold]. In this case, this "decisional" alphabet for each time series has only three letters. The maximal entropy of this alphabet is less than 2 bits. If a decision has to be made for all r time series, we have less than $2r$ bits. Figure 2.9 illustrates the above-mentioned changes of alphabets with different maximal entropy values. The final alphabet ultimately defines the visualization task, while some intermediate alphabets may also capture subtasks in a data analysis and visualization process.

2.3.3 Cost–Benefit Measures for Visualization Processes

From Figure 2.9, one observation that we can make is that there is almost always a reduction of maximal entropy from the original data alphabet to the decisional alphabet. This relates to one of the basic objectives in statistical inference, i.e., to optimize the process between the initial alphabet and the final alphabet with minimal loss of information that is "important" to the decision based on the final alphabet. However, as visualization processes involve both machine-centric and human-centric mappings, it is necessary (i) to optimize both types of mapping in an integrated manner, (ii) to take into account "soft" information that can be introduced by human analysts during the process, and (iii) to consider information loss as part of a cost–benefit analysis.

Let us consider a sequential workflow with L processing steps. As illustrated in Figure 2.8, there are L + 1 alphabets along the workflow. Let $\mathbb{Z}_s$ and $\mathbb{Z}_{s+1}$ be two consecutive alphabets such that

$$F_s : \mathbb{Z}_s \longrightarrow \mathbb{Z}_{s+1}.$$

where F_s is a mapping function, which can be an analytical algorithm that extracts features from data, a visual mapping that transforms data to a visual representation, or a human decision process that selects an outcome from a set of options.

The cost of executing F_s as part of a visualization process can be measured in many ways. Perhaps the most generic cost measure is *energy* since energy would be consumed by a computer to run an algorithm or to create a visualization, and by a human analyst to read data, view visualization, reason about a possible relationship, or make a decision. We denote this generic measurement as a function $C(F_s)$. While measuring energy usage by computers is becoming more practical [121], measuring that of human activities, especially cognitive activities, may not be feasible in most situations. A more convenient measurement is *time*, $C_{time}(F_s)$, which can be considered as an approximation of $C(F_s)$. Another is a monetary measurement of computational costs or employment costs, which represent a subjective approximation from a business perspective. Without loss of generality, we will use $C(F_s)$ as our cost function in this section.

Definition 1 (Alphabet Compression Ratio). As exemplified in Figure 2.9, a mapping function (i.e., a machine or human process) usually facilitates the reduction of the sizes of data alphabets at different stages of data processing though the reduction is not guaranteed. We can measure the level of reduction as the *alphabet compression ratio* (ACR) of a mapping F_s:

$$\Psi_{ACR}(F_s) = \frac{H(Z_{s+1})}{H(Z_s)}, \tag{2.8}$$

where H is the Shannon entropy measure. In a closed machine-centric processing system that meets the condition of a Markov chain, we have $H(Z_s) \geq H(Z_{s+1})$. This is the *data processing inequality* [32]. In such a system, Ψ_{ACR} is a normalized and unitless entropy measure in $[0, 1]$ as first proposed by Golan in [49] (see also [52]). However, Chen and Jänicke pointed out that the Markov chain condition is broken in most visualization processes [25], and further examples were given in [23]. Hence, we do not assume that $H(Z_s) \geq H(Z_{s+1})$ here since F_s can be a human-centric transformation, unless one encodes all possible variants of "soft" information and knowledge in the initial data alphabet.

Meanwhile, given an output of an analytical process, F_s, an analyst will gain an impression about the input. Considering the time-series transformation in Figure 2.9, for example, learning the mean price value for each minute, an analyst may have a conjecture about the 12 original data values. Viewing a visualization of each time-series plot in a resolution of 128 possible values per data point, an analyst may infer, estimate, or guess the time series in its original resolution of 2^{32} possible values per data point. Let us denote an impression about Z_s as a variable Z'_s, which is a result of a mapping G_s such that

$$G_s : \mathbb{Z}_{s+1} \longrightarrow \mathbb{Z}'_s,$$

where $\mathbb{Z}'_s$ is the alphabet of this impression with a probability mass function representing the inferred or guessed probability of each letter in $\mathbb{Z}'_s$. Note that G_s is a reconstruction function, similar to what was discussed in [65]. In most cases, G_s is only a rough approximation of the true inverse function F^{-1}. The difference between such an impression about $\mathbb{Z}'_s$ obtained from observing letters in $\mathbb{Z}_{s+1}$ and the actual $\mathbb{Z}_s$ is defined by Kullback–Leibler divergence (or relative entropy) [32]:

$$D_{KL}(Z'_s||Z_s) = D_{KL}(G_s(Z_{s+1})||Z_s) = \sum_j p(z'_{s,j}) \log \frac{p(z'_{s,j})}{q(z_{s,j})},$$

where $z'_{s,j} \in \mathbb{Z}'_s$, and $z_{s,j} \in \mathbb{Z}_s$, and p and q are two probability mass functions associated with $\mathbb{Z}'_s$ and $\mathbb{Z}_s$ respectively. $D_{KL} = 0$ if and only if $p = q$, and $D_{KL} > 0$ otherwise. Note that D_{KL} is not a metric as it is not symmetric. The definition of D_{KL} is accompanied by a precondition that $q = 0$ implies $p = 0$.

Definition 2 (Potential Distortion Ratio). With the log formula, D_{KL} is also measured in bits. The higher the number of bits is, the further is the deviation of the impression $\mathbb{Z}'_s$ from $\mathbb{Z}_s$. The *potential distortion ratio* (PDR) of a mapping F_s is thus

$$\Psi_{PDR}(F_s) = \frac{D_{KL}(Z'_s||Z_s)}{H(Z_s)}. \tag{2.9}$$

Both $\Psi_{ACR}(F_s)$ and $\Psi_{PDR}(F_s)$ are unitless. They can be used to moderate the cost of executing F_s, i.e., $C(F_s)$. Since $H(Z_{s+1})$ indicates the intrinsic uncertainty of the output alphabet and $D_{KL}(Z'_s||Z_s)$ indicates the uncertainty caused by F_s, the sum of $\Psi_{ACR}(F_s)$ and $\Psi_{PDR}(F_s)$ indicates the level of combined uncertainty in relation to the original uncertainty associated with $\mathbb{Z}_s$.

Definition 3 (Effectual Compression Ratio). The *effectual compression ratio* (ECR) of a mapping F_s from $\mathbb{Z}_s$ to $\mathbb{Z}_{s+1}$ is a measure of the ratio between the uncertainty before a transformation F_s and that after

$$\Psi_{ECR}(F_s) = \frac{H(Z_{s+1}) + D_{KL}(Z'_s||Z_s)}{H(Z_s)} \quad \text{for } H(Z_s) > 0. \tag{2.10}$$

When $H(Z_s) = 0$, it means that variable Z_s has only one probable value, and it is absolutely certain. Hence, the transformation of F_s is unnecessary in the first place. The measure of ECR encapsulates the trade-off between ACR and PDR, since deceasing ACR (i.e., more compressed) often leads to an increase of PDR (i.e., harder to infer Z_s), and vice versa. However, this trade-off is rarely a linear (negative) correlation. Finding the most appropriate trade-off is thus an optimization problem, which is to be further enriched when we incorporate below the cost $C(F_s)$ as another balancing factor.

Definition 4 (Benefit). We can now define the *benefit* of a mapping F_s from Z_s to Z_{s+1} as

$$B(F_s) = H(Z_s) - H(Z_{s+1}) - D_{KL}(Z_s' \| Z_s). \tag{2.11}$$

The unit of this information-theoretic measure is the *bit*. When $B(F_s) = 0$, the transformation does not create any change in the informational structure captured by the entropy. In other words, there is no informational difference between observing variable Z_s and observing Z_{s+1}. When $B(F_s) < 0$, the transformation has introduced more uncertainty, which is undesirable. When $B(F_s) > 0$, the transformation has introduced a positive benefit by reducing the uncertainty. This definition can be related to Shannon's grouping property [32].

Theorem (Generalized Grouping Property). Let X be a variable that is associated with an N-letter alphabet $\mathbb{X}$ and a normalized N-dimensional discrete distribution $p(x), x \in \mathbb{X}$. When we group letters in $\mathbb{X}$ to M subsets, we derive a new variable Y with an M-letter alphabet $\mathbb{Y}$ and a normalized M-dimensional discrete distribution $q(y), y \in \mathbb{Y}$:

$$H(X) = H(Y) + \sum_{k=1}^{M} q(y_k) H_k, \tag{2.12}$$

where H_k is the entropy of the local distribution of the original letters within the kth subset of $\mathbb{X}$. Comparing Equation (2.11) and Equation (2.12), we can see that the last term on the right in Equation (2.12) is replaced with the Kullback–Leibler divergence term in Equation (2.11). The equality in Equation (2.12) is replaced with a measure of difference in Equation (2.11). This is because of the nature of data analysis and visualization. After each transformation F_s, the analyst is likely to infer, estimate or guess the local distribution within each subset, when necessary, from the observation of X in the context of Equation (2.12) or Z_{s+1} in the context of Equation (2.11) in conjunction with some "soft" information and knowledge, as mentioned previously.

Definition 5 (Incremental Cost–Benefit Ratio). The *incremental cost–benefit ratio* (incremental CBR) of a mapping F_s from Z_s to Z_{s+1} is thus defined as the ratio between benefit $B(F_s)$ and cost $C(F_s)$:

$$\Upsilon(F_s) = \frac{B(F_s)}{C(F_s)} = \frac{H(Z_s) - H(Z_{s+1}) - D_{KL}(Z_s' \| Z_s)}{C(F_s)}. \tag{2.13}$$

Note that we used cost as the denominator because (i) the benefit can be zero, while the cost of transformation cannot be zero as long as there is an action of transformation; and (ii) it is better to associate a larger value with a more cost-beneficial meaning.

At each transformation s, if one changes the method (e.g., a visual design or an analytical algorithm), the change will likely affect the amount of alphabet compression, $H(Z_s) - H(Z_{s+1})$, potential distortion $D_{KL}(Z'_s||Z_s)$, and cost $C(F_s)$. For example, as illustrated in the top-left block in Figure 2.10, one may wish to increase alphabet compression by (a) using a more abstract visual representation or (b) rendering a visualization at a lower resolution. Approach (a) may increase the cost while reducing (or increasing) the potential distortion. Approach (b) may reduce the cost while increasing the potential distortion. Hence, discovering the best method is an optimization process.

Furthermore, the change of a method at one transformation may likely trigger subsequent changes in the succeeding transformation. This cascading effect is illustrated in Figure 2.10.

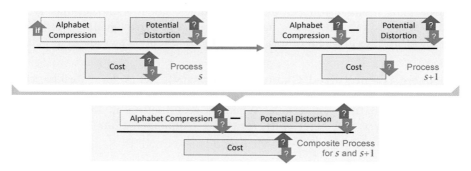

Figure 2.10 Making changes to a transformation may result in changes in the amount of alphabet compression, potential distortion, and cost within the transformation (top left). It may also have cascading effects on the succeeding transformation (top right), and combined effects on the composite transformation (below) [24].

For example, if one uses a visualization with a high alphabet compression ratio at transformation F_s, a human observer may be able to observe a different set of features, resulting in a change of the feature alphabet and thus the alphabet compression ratio in the following transformation F_{s+1}. Even when the feature alphabet remains the same, the human observer may (or may not) recognize various multivariate features more accurately (i.e., less potential distortion) or speedily (i.e., less cost). Hence it is necessary to optimize the combined cost–benefit ratio of a composite transformation.

Given a set of cascading mapping functions, $F_1, F_2, \ldots, F_L$, which transform alphabets from $\mathbb{Z}_1$ to $\mathbb{Z}_{L+1}$, we can simply add up their costs and benefits as

$$C_{total} = \sum_{s=1}^{L} C(F_s)$$

$$B_{total} = \sum_{s=1}^{L} E(F_s) = H(Z_1) - H(Z_{L+1}) - \sum_{s=1}^{L} D_{KL}(Z_s'||Z_s).$$

The *overall cost–benefit ratio* (overall CBR) is thus B_{total}/C_{total}.

For workflows containing parallel mappings, the merging of CBR at a joint partly depends on the semantics of the cost and benefit measures. If we are concerned about the energy, or monetary cost, the simple summation of cost measures arrived at a joint makes sense. If we are concerned about the time taken, we may compute the maximum cost at a joint. If all parallel branches arriving at a joint contain only machine-centric processes, the benefit is capped by the entropy at the beginning of the branching-out. The combined benefit can be estimated by taking into account the mutual information between the arriving alphabets. When these parallel branches involve human-centric processing, "soft" information will be added into the process. The combined benefit can be estimated in the range between the maximum and the summation of the arriving benefit measures.

Here we largely focus on the workflows for conducting data analysis and visualization. The above formulation of cost–benefit analysis can be extended to include the cost of development and maintenance.

2.3.4 Examples of Cost–Benefit Analysis

There are numerous kinds of visualization tasks, which typically depend on applications, datasets, and users. Conceptually, we can categorize visualization tasks into four different levels [24], reflecting the complexity of visualization tasks from the perspective of analysts.

- *Level 1: Disseminative visualization.* Visualization is a presentational aid for disseminating information or insight to others. The analyst who created the visualization does not have a question about the data, except for informing others: "This is A!" where A may be a fact, a piece of information, an understanding, etc. At this level, the complexity for the analyst to obtain an answer about the data is $O(1)$. Here we use the big O notation in algorithm and complexity analysis. However, instead of measuring the complexity of computation or storage costs, we focus on the search space for answers in performing a visualization task.

- *Level 2: Observational visualization.* Visualization is an operational aid that enables intuitive and/or speedy observation of captured data. It is often a part of routine operations of an analyst, and the questions to be answered may typically be "What has happened?" "When and where did A, B, C, etc., happen?" At this level, the observation is usually sequential, and thus the complexity is generally $O(n)$, where n is the number of data objects.

- *Level 3: Analytical visualization.* Visualization is an investigative aid for examining and understanding complex relationships (e.g., correlation, association, causality, contradiction). The questions to be answered are typically "What does A relate to and why?" Given n data objects, the number of possible k-relationships among these data objects is at the level of $O(n^k)$ ($k \geq 2$). For a small n, it may be feasible to examine all k-relationships using observational visualization. When n increases, it becomes necessary to use analytical models to prioritize the analyst's investigative effort. Most visual analytics processes reported in the recent literature operate at this level.

- *Level 4: Model-developmental visualization.* Visualization is a developmental aid for improving existing models, methods, algorithms, and systems, as well as for creating new ones. The questions to be answered are typically "How does A lead to B?" and "What are the exact steps from A to B?" If a model has n parameters and each parameter may take k values, there are a total of k^n combinations. In terms of complexity, this is $O(k^n)$. If a model has n distinct algorithmic steps, the complexity of their ordering is $O(n!)$. Model-developmental visualization is a great challenge in the field of visualization.

Hence the levels correspond to the questions to be asked and the complexity of the space of optional answers. Tasks of different levels can be featured in different workflows individually or in an integrated manner. In the following subsections, we consider several successful visualization processes in the literature. In particular, we first examine an example of interaction techniques since interaction plays a significant role in breaking the conditions of data processing inequality. We then examine one example for each of the four levels of visualization.

We analyze their cost–benefit ratios in comparison with possible alternative processes. The comparison also serves as initial validation of the information-theoretic measures proposed in the previous section.

2.3.4.1 Interaction in Visualization

In data analysis and visualization, human–computer interaction plays a significant role in breaking the condition of the *data processing inequality* [25]. It enables human analysts to introduce "soft" information and knowledge into such a process. Here we consider that the initial alphabet at the beginning of the process represents "hard" data, e.g., $\mathbb{Z}_1$ for representing variants of r time series. The "soft" information and knowledge is "external" to the process, which can no longer be a closed system.

Interaction has been studied extensively in the context of visualization (e.g., [28, 125, 141, 149]). One of the commonly used forms of interaction is "overview first, zoom, and details on demand" [109]. It may feature several types of actions, including *select, explore, abstract/elaborate* and *filter* as defined by Yi et al. [149]. Figure 2.11 illustrates such a process with the abstract notion presented in Section 2.1.1. Under an information-theoretic framework, the input and output of each transformation are considered in a holistic manner, i.e., as alphabets (e.g., all variants of images that may be displayed in a context) rather than individual letters (e.g., an image).

One may imagine a very large image (or map) as an instance of alphabet $\mathbb{Z}_s$ at the top of Figure 2.11. An interactive system first presents viewers with an overview, which is an instance of alphabet $\mathbb{Z}_{s+1}$. A viewer may select a part of the overview and apply a zoom-in operation. The detailed view at that location is an instance of alphabet $\mathbb{Z}_{s+2}$, which is a subset of $\mathbb{Z}_s$. From this detailed view, the viewer may choose to explore to a nearby location, and so on. At some stage, the viewer decides to finish the exploration and makes up his/her mind about something based on the overview and parts of the full image (or map) representation that has been explored so far. The examples of "soft" information and knowledge in this case may include how important the individual subsets of $\mathbb{Z}_s$ are to the viewer, and which direction of exploration from one subset to the next is more promising.

Recall the example illustrated in Figure 2.7 and discussed in Section 2.2, where two different overview techniques for flow visualization were used to illustrate the optimization based on mutual information I. Although the concept of cost–benefit analysis was not featured in Section 2.2, the idea of optimization in that example can be translated to an information-theoretic measure of potential distortion.

Given a transformation $F_s : \mathbb{Z}_s \longrightarrow \mathbb{Z}_{s+1}$, mutual information $I(Z_s; Z_{s+1})$ measures the amount of uncertainty shared between the input and output. Knowing all about Z_{s+1} does not normally imply knowing all about Z_s, i.e., $I(Z_s; Z_{s+1}) \leq H(Z_s)$. Hence, even if one has full knowledge about Z_{s+1}, the uncertainty about Z_s is at least

$$H(Z_s) - I(Z_s; Z_{s+1}).$$

This measure can also be used as a measure of potential distortion with an assumption that a perfect inverse mapping F_s^{-1} from $\mathbb{Z}_{s+1}$ to $\mathbb{Z}_s$ can infer all mutual information, but no more than that. Hence, we can rewrite Equation (2.13) as

$$\Upsilon(F_s) = \frac{B(F_s)}{C(F_s)} = \frac{I(Z_s; Z_{s+1}) - H(Z_{s+1})}{C(F_s)}. \tag{2.14}$$

This criterion is conceptually consistent with Equation (2.13), though it offers a different measure. Equation (2.14) is thus a mathematical representation of what was outlined in Section 2.2.

In Figure 2.11, the transformations following $\mathbb{Z}_{s+1}$ are all human-centric

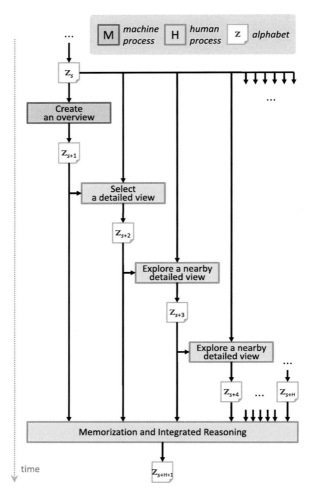

Figure 2.11 Interaction is one of the means for introducing "soft" information into a visualization process. This figure shows a sequence of interactions for overview first and details on demand. At some stage of a visualization process, the system receives a large detailed visual representation $z \in \mathbb{Z}_s$. It creates an overview. A viewer selects a part of the overview and requests a detailed view, which is part of z. From this detailed view, the viewer explores a few nearby detailed views. At some stage, the viewer decides to finish the exploration and makes up his/her mind about something based on the overview and partial observation of the detailed visual representation. At each step, all possible valid inputs and outputs of a transformation are letters of an alphabet.

processes. We can observe that these transformations will incur costs such as cognitive effort and time for interaction. One important consideration is the *prior knowledge* about $\mathbb{Z}_s$, i.e., how much information is already known to the viewers, and how much is uncertain. Considering the same flow visualization example as in [25], one may estimate the following:

- How likely is it that viewers know that $\mathbb{Z}_s$ are texture-based representations of vector fields?

- How confident are viewers about the correctness of the feature-extraction technique used to create an overview?

- How long have the viewers been working on the simulation model that generates the vector fields being visualized?

Such inference can be translated to an estimation about the term $D_{KL}(Z'_s||Z_s)$ in Equations 2.10 and 2.13. Some further optimization can be often implemented on top of the basic form of *overview first and details on demand*. In many scenarios, we often observe that an experienced viewer may find step-by-step zoom operations frustrating, as the viewer knows exactly where the interesting part of a detailed representation is. For example, in flow simulation, scientists often work on the same simulation problem for months, and have a good mental overview about $\mathbb{Z}_s$. In such a case, when the interactive visualization system has a fast track for reaching a specific detailed view (e.g., the last location visited), it reduces the cost of step-by-step zoom operations. However, this approach may not be applicable to an online map system, where each search session is likely to be initiated for a new search task.

2.3.4.2 Disseminative Visualization

The history of the time-series plot can be traced back more than a millennium. If success is measured by usage, it is undoubtedly one of the most successful visual representations. However, its display space utilization is rather poor in comparison with a binary digits view [25]. Figure 2.12 shows two such representations that are used as disseminative visualization for a scenario in Figure 2.9. The dataset being displayed is a time series with 60 data points, i.e., an instance of $\mathbb{Z}_2$ in Figure 2.9. Assume that the value of this particular share has been largely moving between 100 and 200 cents. Hence the entropy of $Z_{a,1} = Z_{b,1}$ is estimated to be about 420 bits, significantly below the maximal entropy of the data representation.

The binary digits view uses a 2×2 pixel-block per digit, and requires 32×60 blocks (7,680 pixels) for the plotting canvas. Using the same number of pixels, 128×60, the time-series plot is an instance of $\mathbb{Z}_{a,2}$. During dissemination, the presenter (or analyst) points out "stable" and "rise" features to a viewer (or client), suggesting a decision "to hold." The overall CBRs for the two pipelines in Figure 2.12 are

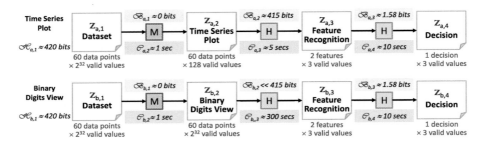

Figure 2.12 Comparison between a time-series plot and a binary digits view for disseminative visualization. The same legend in Figure 2.9 applies. The estimated benefit and cost values here are based on heuristic reasoning and for an illustrative purpose. For example, for $B_{a,2}$, we consider two feature variables [stable, uneven, volatile] and [rise, fall, flat]. Hence the maximal entropy of $Z_{a,3}$ is about 3.17 bits. As the D_{KL} term for $B_{a,2}$ will indicate some uncertainty, the estimated benefit is $420 - 3.17 - D_{KL}(Z'_{a,2}||Z_{a,2}) \approx 415$ bits. Meanwhile, D_{KL} for $B_{b,2}$ is much higher [24].

$$\Upsilon_{plot} = \sum_{j=1}^{3} \frac{H(Z_{a,j+1}) + D_{KL}(Z'_{a,j}||Z_{a,j})}{C(F_{a,j})} \tag{2.15}$$

$$\Upsilon_{binary} = \sum_{j=1}^{3} \frac{H(Z_{b,j+1}) + D_{KL}(Z'_{b,j}||Z_{b,j})}{C(F_{b,j})}. \tag{2.16}$$

To the presenter, the decision "to hold" has already been made, and the total CBR would be zero for either workflow. For a viewer unfamiliar with binary representations, the binary digits view is almost undecipherable. For a pair of untrained eyes, recognizing features such as "stable" and "rise" would take a while. The inverse mapping from the features pointed out by the presenter is also rather uncertain, hence a high value for the D_{KL} term in $B_{b,2}$. The binary digits view thereby incurs a huge cost at the feature recognition step, while bringing lower benefit. This mathematically explains the merits of the time-series plot over a spatially compact binary digits view.

2.3.4.3 Observational Visualization

The example in Section 2.3.4.2 can also be considered in the context of observational visualization, where an analyst creates a visualization for him/herself. Similar abstract reasoning and step-by-step inference can be carried out, just as in the previous example, likely for a much larger input data alphabet (e.g., with r time series and t hours).

Let us consider a different example of observational visualization. Legg et al. reported an application of visualization in sports [77]. The Welsh Rugby Union required a visualization system for in-match and post-match analysis. One of the visualization tasks was to summarize events in a match, facilitating external memorization. The input datasets are typically in the form of videos including data streams during a match, and can be generalized to include direct viewing of a match in real time. The alphabet is thus huge. The objective for supporting external memorization is to avoid watching the same videos repeatedly. Especially during a half-time interval, coaches and players cannot afford much time to watch videos.

The workflow can be coarsely divided into three major transformations, namely F_a: transforming real-world visual data to events data, F_b: transforming events data to visualization, and F_c: transforming observations to judgments and decisions. Clearly, transformation F_c should be performed by coaches and other experts. For transformation F_a, two options were considered: $F_{a,1}$ for computers to detect events, and $F_{a,2}$ for humans to detect events. For transformation F_b, two options were considered: $F_{b,1}$ statistical graphics, and $F_{b,2}$ glyph-based event visualization. For $F_{a,1}$ and $F_{a,2}$, the letters of the output alphabet are multivariate data objects describing what type of event, when and where it happens, and the players involved. This alphabet is much smaller than the input alphabet for real-world visual data.

The team did not find any suitable computer vision techniques that could be used to detect events and generate the corresponding data objects in this application. The accuracy of available techniques was too low, hence the D_{KL} term for $F_{a,1}$ yields a high level of uncertainty. Using a video annotation system, an experienced sports analyst can generate more accurate event data during or after a match. For an 80-minute rugby match, the number of data objects generated is usually in hundreds and sometimes in thousands. Hence, statistics can be obtained, and then visualized using statistical graphics. However, it is difficult for coaches to make decisions based on statistical graphics, as it is difficult to connect statistics with episodic memory about events. Such a difficulty corresponds to a high level of uncertainty resulting from the D_{KL} term for $F_{b,1}$. On the other hand, the direct depiction of events using glyphs can stimulate episodic memory much better, yielding a much lower-level uncertainty in the D_{KL} term for $F_{b,2}$. The team implemented $F_{a,2}$ and $F_{b,2}$ transformations as reported in [77], while $F_{b,1}$ was also available for other tasks.

2.3.4.4 Analytical Visualization

Oelke et al. studied a text analysis problem using visual analytics [89]. They considered a range of machine-centric and human-centric transformations in evaluating document readability. For example, the former includes 141 text feature variables and their combinations. The latter includes four representations at three different levels of detail. Since different combinations of machine-centric and human-centric transformations correspond to different visual analytics pipelines, their work can be seen as an optimization effort. Through experimentation and analysis, they confirmed the need for enabling analysts to observe details at the sentence or block levels. Over-aggregation (e.g., assigning a readability score to each document) is not cost beneficial, as the trade-off between the alphabet compression ratio (ACR) and the potential distortion ratio (PDR) is in favor of the PDR.

2.3.4.5 Model-Developmental Visualization

In [118], Tam et al. compared a visualization technique and a machine-learning technique in generating a decision tree as a model for expression classification. The input to this model development exercise is a set of annotated videos, each of which records one of four expressions [anger, surprise, sadness, smile]. The output is a decision tree that is to be used to classify videos automatically with reasonable accuracy. It is thus a *data analysis and visualization process* for creating a *data analysis model.* Although this sounds like a conundrum, it fits well within the scope of visualization. Tam et al. approached this problem through a series of transformations. The first transformation F_a identifies 14 different facial features in each video, and records its temporal changes using a geometric or texture measurement. This results in 14 different alphabets of time series. The second transformation F_b characterizes each time series using 23 different parameters. This results in a total of $322 = 14 \times 23$ variables. At the end of the second transformation, each video becomes a 322-variate data object.

For the visualization-based pipeline, the third transformation $F_{c,1}$ generates a parallel coordinate plot with 322 axes. This is followed by the fourth transformation $F_{d,1}$, where two researchers laid the big plot on the floor and spent a few hours selecting the appropriate variables for constructing a decision tree. For the machine-learning-based pipeline, the team used a public-domain tool, C4.5, as the third transformation $F_{c,2}$, which generates a decision tree from a multivariate dataset automatically.

In terms of time cost, transformation $F_{c,2}$ took much less time than transformations $F_{c,1}$ and $F_{d,1}$ together. In terms of performance, the decision tree created by $F_{c,1}$ and $F_{d,1}$ was found slightly more accurate than that resulting from $F_{c,2}$. From further analysis, they learned that (i) handling real values has been a challenge in automatic generation of decision trees; (ii) the two researchers did not rely solely on the parallel coordinates plot to choose variables, and their "soft" knowledge about the underlying techniques used in transformations F_a and F_b also contributed to the selection. Such "soft" knowledge reduces the uncertainty expressed by the D_{KL} term in Equation 2.11. This example demonstrates the important role of visualization in model development.

2.4 INFORMATION-THEORETIC LINKS: VISUALIZATION AND PERCEPTION

2.4.1 Visual Multiplexing

2.4.1.1 Multiplexing in Communication

In the history of communication, *multiplexing* was introduced along with telegraph systems in the 19th century [124]. The concept, which is underpinned by information theory, enables efficient use of a transmission system, by allowing a number of signals at a lower bit rate to share a single higher-bit-rate transmission medium [113]. By improving the utilization of channel capacity, multiplexing brings about a huge amount of benefits such as cost effectiveness and convenience in sharing a common medium. In a multiplexing system, a *multiplexer* (mux) is an encoder that combines multiple input signals and forwards the combined signal onto a transmission medium. A *demultiplexer* (demux) is a decoder that takes a signal from a shared medium and separates it into multiple signals. The commonly used multiplexing schemes are as follows:

- *Frequency-division multiplexing* (FDM). It divides the transmission frequency range into narrow bands and is widely used in radio, television, telephone, and satellite carrier systems [6] (p.212).

- *Time-division multiplexing* (TDM). It provides a user with the full channel capacity but divides the channel usage into time slots [6] (p.215). It is commonly used in digital communication.

- *Space-division multiplexing* (SDM). This is observable in wired communication systems, where a cable consists of a bundle of many wires carrying different signals.

- *Code-division multiplexing* (CDM). It enables different signals encoded in different coding schemes to share a single transmission line at the same time over the same frequency band [114].

2.4.1.2 Multiplexing in Visualization

Intriguingly, visualization also exhibits a variety of multiplexing phenomena. Consider 2D media where visualizations are displayed. Such media may encompass many types of computer displays, projection screens, and printed media. (Our consideration does not include any stereoscopic or holographic devices, which will make an interesting topic for future research.)

Definition. Let p be a point on a medium, x be a valid value of a data variable, and c be a visual signal in a visual channel. Given a multivariate data object, $X = \langle x_1, x_2, \ldots, x_k \rangle (k > 1)$, associated with a location p, the goal of a *visual multiplexer* is to map each data value x_i to a visual signal c_i, and deliver $\langle c_1, c_2, \ldots, c_k \rangle$ through the corresponding visual channels.

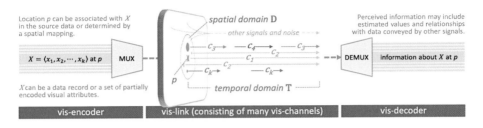

Figure 2.13 This illustration shows a multivariate data object X associated with a location p in a visualization. The *visual encoder* maps the data values of X to different visual signals that may be partitioned spatially and temporally and may use different amounts of spatial or temporal bandwidth. The *visual decoder* attempts to recover information about X, for example, by estimating its data values and determining its relationships with data conveyed by other signals. The mux and demux are part of the *vis-encoder* and *vis-decoder* respectively. Similar to the terminology in communication, the connection between the mux and demux is referred to as a *vis-link*, which consists of multiple visual channels (cf. Figure 2.2) [26].

It is not essential for these visual signals be located at p and they can be distributed in a spatial domain **D** surrounding p. It is not necessary for these visual signals to be present at the same time either and they can be delivered over a short time interval **T**. Different signals may be mixed in display. On the other hand, the goal of a *visual demultiplexer* is to obtain information about $\langle x_1, x_2, \ldots, x_k \rangle$ at p by observing and interpreting visual signals depicted in **D** during **T**. Figure 2.13 illustrates the concept of visual multiplexing. In the following discussions, we will simply refer to the visual multiplexer and demultiplexer as *mux* and *demux*, respectively, as in communication. One major departure from a communication system is that a demux in a visualization pipeline is not a technical device, but a human-centered function that relies on the perception and cognition of a viewer.

Visual multiplexing may take several different forms as illustrated in Figure 2.14. The process for deriving this categorization is given in the supplementary materials. There are two basic forms of multiplexing, conceptually similar to SDM and TDM in communication. Almost all visual designs in visualization involve partitioning a 2D medium into different regions, and delivering multiple visual signals simultaneously through these regions, which can be fine grained at the level of pixels or ink dots. Figure 2.14(a) illustrates a coarse-grained space division with a pixel-based visualization, and a fine-grained space division with an elevation map. Animation is a basic form of time-division multiplexing. Theoretically, each pixel in every frame can act as a visual channel by transmitting a signal independently. In practice, it is well understood that a human viewer would not normally obtain all information at such a fine-grained level. Figure 2.14(b) shows a simple example, where signals for different data values (rain, temperature, wind speed, and direction) can be transmitted and received in a short animation.

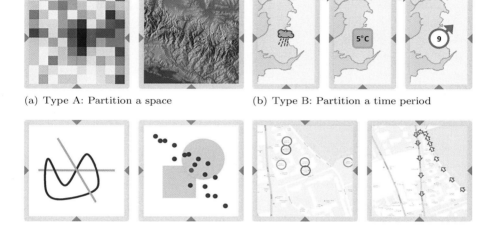

(a) Type A: Partition a space

(b) Type B: Partition a time period

(c) Type C: Introduce partial occlusion

(d) Type D: Use "hollow" visual channels

Figure 2.14 Ten different types of visual multiplexing. While multiplexing may occur in several places in each illustration, we focus on a specific point p indicated by arrows on the bounding box. The first four are shown here and six more are shown in Figures 2.15 and 2.16 [26].

Let us now consider more complex forms of multiplexing by focusing on a single point p, marked by grey arrows on the bounding box of each illustration in Figure 2.14. One crude but often effective means of multiplexing is simply for the mux to plot all k visual channels at the same position p, allowing the last one to be displayed to occlude all other visual channels at p. Figure 2.14(c) shows two illustrations, where the viewer can easily perceive three pieces of information at the center of the square due to some gestalt effects, e.g., in these two cases, *continuity*, *similarity*, and *proximity*. Depending on the availability of a priori knowledge about the possible types of shapes in a visualization, the opaque occlusion may result in some uncertainty. For example, one may be uncertain about whether the orange rectangle does reach p, the center of the square in the right illustration.

Such uncertainty can be reduced by replacing filled shapes with outlines in Figure 2.14(c). Some visual channels do not necessarily demand the use of

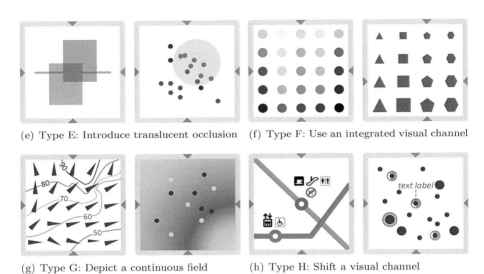

(e) Type E: Introduce translucent occlusion (f) Type F: Use an integrated visual channel

(g) Type G: Depict a continuous field (h) Type H: Shift a visual channel

Figure 2.15 Ten different types of visual multiplexing. Continuing from Figure 2.14, four more types of visual multiplexing are shown here and two more are shown in Figure 2.16 [26].

p in the medium in order to deliver information for p. As illustrated in Figure 2.14(d), visual channels such as *shape* and *size* can have an hollow interior, hence freeing the optical properties of p for delivering other signals. Some topological and relational channels, such as distance, closure, and density, may naturally leave a fair amount of space unblocked. In the right illustration of (d), the distance between arrows is a visual channel for depicting speed. At p, the center of the illustration, the demux delivers the speed information, while revealing the map information at p.

As illustrated in Figure 2.15(e), translucency allows some "non-hollow" visual channels to benefit in a way similar to a "hollow" visual channel. In addition to gestalt effects, the demux involves some cognitive reasoning, e.g., in these two cases, about the colors. For the left illustration, the demux has to determine if the purple square in the middle is an independent shape or the blending of the blue and red rectangles. For the right illustration, the demux has to reason about whether there are 2, 3, or 4 types of dots, since there are 4 differently colored dots. The knowledge about color blending thus becomes an "algorithmic" aspect of the demux.

At location p, translucency introduces an integration of different optical signals. In Figure 2.15(e), the demux attempts to differentiate one signal from another. In many other cases, the demux has to separate different signals at p more or less depending only on the integrated signal at p. Figure 2.15(f) illustrates two integrated visual channels, on the left, hue (x-axis) and luminance

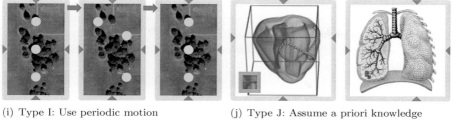

(i) Type I: Use periodic motion   (j) Type J: Assume a priori knowledge

(k) Type J (continued): Acquired knowledge   (l) Type J (continued): Visual language

Figure 2.16 Ten different types of visual multiplexing. Continuing from Figures 2.14 and 2.15, two more types of visual multiplexing are shown here. For Type J (model-based), three sets of illustrations are shown [26].

(y-axis); and on the right, shape or number of edges (x-axis) and size (y-axis). Although it is known that integrated channels have some shortcomings [108], nevertheless, they provide a means of multiplexing.

In visualization, a 2D field, which assumes continuous coverage of every point in a spatial domain, is usually depicted by using discretely placed objects. For example, the left illustration of Figure 2.15(g) shows a height field with a set of contour lines and a vector field that is depicted using regularly spaced arrows. The demux can establish two pieces of information at the center p (and all other points) in a way that very much resembles the algorithmic function of bilinear interpolation. With the example on the right, similarly, the demux can infer the possible colors under the dots from the nearby colors that are not occluded.

In other cases, the demux may use the "nearest neighbor" interpolation, as illustrated in Figure 2.15(h). The various signs shown in the left illustration are individual visual channels that convey different information associated with the positions marked by hollow circles. At the center location, p, despite the fact that the mux has deliberately displaced the 4 signs away from p, the demux has no difficulty in translating the information back to p.

In Figure 2.16(i), displacement is also used, but in a temporal manner rather than as a "permanent" feature (cf. (h) Figure 2.15). The three yellow dots move in different directions periodically, thereby encoding three vectors.

Because of the motion, none of the pixels in the aerial photograph are totally occluded within a time window. Conceptually, this may be considered similar to the use of "hollow" visual channels in Figure 2.14(d), except the "hollowing out" takes place in the temporal domain $\mathbf{T}$. Note that this is not quite the same as the basic time-division multiplexing in Figure 2.14(b), where signals for different data values are delivered in an ordered manner.

While for Types A-I, the demux obtains some information at p from the surrounding domain $\mathbf{D}$ or a time interval $\mathbf{T}$ with little cognitive effort, often there are also cases where the demux relies on a known model about what is expected to be seen and how to see it. For example, Figure 2.16(j) shows a volume-rendered object and a medical illustration. Once aware of the fact that the object on the left is a heart, the demux can perceive several surfaces along the ray passing through the center point p, though the actual quality of such signals is very poor (cf. the bottom-left inset). The cognitive reasoning relies on the a priori knowledge about the structure of a heart, such as the number of chambers. Similarly, with the right illustration, the demux can make use of the a priori knowledge of the relationship between two lungs to see multiple pieces of information displayed on the left and right lungs by applying a symmetry rule.

Specific knowledge for visual multiplexing can be acquired. Figure 2.16(k) shows two simple examples. In visualizing schematic logic diagrams, there is often uncertainty about connectivity between two crossing lines if one relies solely on gestalt effects (e.g., Type C). It is common to introduce a specialized depiction, such as a circle for a joint, and a blank interval or distortion for unconnected lines. Similarly, a solid yellow box that would cause severe occlusion can be replaced by a wireframe mesh, a metaphoric illustration, or a piecewise deformed box. Such knowledge may not be intuitive at the beginning, but can be learned quickly. Note that the knowledge-based depictions in Figure 2.16(k) make use of other types of multiplexing, including Types C, D, and H.

In some applications, the knowledge required for decoding complex multiplexing can be formalized into a *visual language*. To illustrate this, Figure 2.16(l) shows a simple glyph design with two graphical components. It consists of a number of visual channels, which can be used to encode different data variables. The relatively independent visual channels include shape, size, color, the aspect ratio of each outline, and the relative location of the two outlines in each glyph. Each visual channel is a hollow channel (i.e., Type D). In other words, they can encode multivariate data objects without causing much occlusion, for instance, to the continuous field below.

In their work on an information-theoretic framework [25], Chen and Jänicke suggested that multiplexing may be relevant to comparative visualization, volume rendering, and multifield visualization. The illustrative examples in Figure 2.15(g,h) characterize typical multivariate, multifield, and multidimensional visualization applications, while Figure 2.16(j) exemplifies volume visualization.

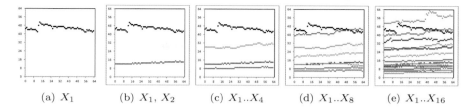

(a) X_1 (b) X_1, X_2 (c) $X_1..X_4$ (d) $X_1..X_8$ (e) $X_1..X_{16}$

Figure 2.17 Uniplexed and multiplexed dot plots for time-series visualization. Plot (a) shows a uniplexed time series that is an instance of a data variable X_1. Plot (b) shows two multiplexed time series, one is an instance of a data variable X_1 and another is of X_2. Plot (c) shows four multiplexed time series. Plot (d) shows eight multiplexed time series. Plot (e) shows sixteen multiplexed time series, each of which is an instance of an independent data variable $X_i, i = 1, 2, \ldots, 16$. Although the original data was obtained from DataMarket.com, it is not intended here to visualize that dataset. In order to illustrate the calculation of information-theoretic measures, the data has been mapped to the integer value range between 0 and 63 [26].

2.4.1.3 Multiplexing and Information-Theoretic Measures

Multiplexing is an integral component of modern communication technologies. This suggests that visual multiplexing may also be an integral part of visualization. One can find a wealth of evidence from both perception literature and existing works in visualization to confirm the existence of visual multiplexing as a common phenomenon [26]. As important as empirical evidence, the existence of visual multiplexing can also be explained using information theory.

Let us revisit the measurement of *display space utilization* (DSU) introduced in Section 2.1.2:

$$\text{Display space utilization (DSU)} = \frac{V(G)}{D},$$

where D is the *display space capacity*, and $V(G)$ is the *visualization capacity* of a graphical mapping G. Using a time-series plot as an example, they showed that the DSU in that particular condition is 0.03125. In other words, from an information-theoretic perspective, only 3.1% of D is utilized by that time-series plot in [25]. Hence visual multiplexing merely provides a means for making more efficient use of the apparently wasteful display space capacity. Also using time-series plots as an example, we demonstrate how this is achieved in terms of DSU.

Figure 2.17 shows five simple time-series plots. Assume that the graph plotting area (i.e., excluding axis regions) in each plot is given as 64×64 pixels.

The medium (or device) used is in 24-bit color. Hence $D = 24 \times 2^{12} = 3 \times 2^{15}$ bits.

In Figure 2.17(a), a dot plot shows one time series with 64 independent samples, each of which has an integer value range between 0 and 63. We choose the dot plot as it is more intuitive than a line graph for illustrating the calculation of various probabilistic measures. Samples are taken at a regular temporal step, so the dot plot displays 64 pixels corresponding to the data samples. The probability mass function of each sampling value is independent and identically distributed, i.e., $p = 1/64$. Let X_1 denote the data variable encompassing all time series with 64 samples in the integer range $[0, 63]$. The time series in Figure 2.17(a) is just an instance of X_1. It is not difficult to calculate the entropy of X_1, yielding $H(X_1) = 384$ bits. Using the same reasoning as [25], we obtain that $V(G) = H(X_1) = 384$ bits. Therefore DSU $= 384/(3 \times 2^{15}) = 0.0039$. Thus, D is very much underutilized and has more than 99% spare capacity.

Let X_2 be the data variable for a second time series with the same numerical and probabilistic constraints. The entropy of X_2 is also 384 bits. As it has been commonly done in practice, we can multiplex two time series in the same plot area. As shown in Figure 2.17(b), the two instances of X_1 and X_2 happen to occupy different parts of the plot area without any overlapping. This can be considered as spatial multiplexing (i.e., Type A). If we use this visual mapping G for any arbitrary pair of time series from X_1 and X_2, we will nearly double the visualization capacity $V(G)$, such that $V(G) = H(X_1) + H(X_2) - \epsilon_{1,2}$, where $\epsilon_{1,2}$ is the average information loss due to overlapping. As discussed in the previous section, gestalt effects (i.e., Type C) can alleviate a fair amount of the perceptual difficulties caused by overlapping unless that one series is largely occluded by another. In fact, the probability of such a situation is rather low. In our example, at each of the 64 sampling positions, the probability of two series coinciding is $p = 1/64$. Considering all 64 positions, the probability of k coincident positions is

$$Q(k) = \binom{64}{k} p^k (1-p)^{64-k} = \frac{64!}{k!(64-k)!} p^k (1-p)^{64-k}.$$

Hence, the probability that two time series coincide at 4 or fewer positions is $\sum_{k=0}^{4} Q(k) = 0.997$. We are reasonably certain that gestalt effects can comfortably help separate two time series coinciding at 4 or fewer positions (out of 64).

From Figures 2.17(d,e), we can see that the situation deteriorates when the amount of multiplexing increases. Most people have found, or would find, that visualizing 10 or more time series in the same plot is not easy. Even if we ignore the information loss ϵ, multiplexing 10 time series would result in DSU $= 0.039$. The utilization of D is more efficient, but it is nowhere near the capacity limit.

Some may wonder where the spare capacity of D comes from for visualization that makes full use of every pixels, e.g., the right-side illustrations in Figures 2.15(g) and 2.16(k,l). Let us consider that Y is the data variable encompassing all background heatmaps in these images. In comparison with the data variable Ω for all possible images that can be displayed on the same display, Y is usually tiny because most images (e.g., portraits) are not probable in that specific context. Hence the entropy $V(G) = H(Y)$ is much smaller than $H(\Omega) = D$, therefore DSU $\ll 1$. The substantially underutilized channel capacity makes it impossible to introduce visual multiplexing.

2.4.2 Information-Theoretic Quality Metrics

The goal of a visualization image is to communicate information about the underlying data. Many creators of visualizations base their choice of techniques, parameter settings, and styles on an intuitive understanding, experience, and personal preferences, which results in a large variety of possible visualizations using different methods and parameter settings. It is thereby desirable to judge the quality of a visualization image, to compare different visualization techniques, and to optimize a parameter setting.

Dictionaries define *quality* in general as a grade of excellence or superiority. As for visualizations, a technique is commonly considered to be superior to another one, if questions concerning the underlying data can be answered more easily, faster, or more accurately. Often a specific aspect of a visualization's quality is assessed by using an empirical study (or studies) where participants are asked to perform visualization tasks and quantitative performance data are then collected and analyzed to yield some conclusions about the quality.

While user studies are very useful to evaluate some fundamental characteristics of a technique, it is not possible to conduct an empirical study for each individual visualization every time it is created. The quality of visualization images depends on main factors, including the domain-specific requirements, the user's needs and expectations, the source dataset and the techniques used. Hence, it is desirable to provide users with alternative means to measure visualization quality. Ideally, such a measure is generic, readily available, and easily applicable to many types of visualization. The requirements for such a quality metric include the following:

- It must be a verifiable measure stated in quantitative terms.

- It allows and encourages comparison between visualization results generated using different control parameters.

- It provides a computer-generated measure complementary to results from empirical studies as it is not practical to conduct a user study for each visualization image.

Meanwhile, it is unrealistic to expect a single quality metric to adequately serve all needs. Combining different quality measures might result in more comprehensive feedback on the quality of a visualization that helps create correct, meaningful, easy-to-understand, and aesthetic visualizations.

Researchers in visualization have proposed many general quality metrics. Among these, some are based on information-theoretic measures. Two of the most extensive uses of these measures are view selection and view optimization, which will be detailed in later chapters. Information-theoretic measures have also been used to measure levels of privacy and utility in privacy-preserving visualization [37]. In the following three subsections, we describe an information-theoretic metric, proposed by Legg et al. [79], for measuring the differentiability between glyphs.

2.4.2.1 Differentiability of Glyphs

Glyph-based visualization [9, 140] is a common form of visual design where some data records are depicted by pre-defined visual objects, which are called *glyphs*. Glyph-based visualizations are ubiquitous in modern life since they make excellent use of the human ability to learn abstract and metaphoric representations to facilitate instantaneous recognition and understanding. Glyphs can be used to encode variables of different data types, categorical (e.g., [77, 82]) as well as numerical (e.g., [42, 72]). However, glyphs are typically small, and are often designed with a high-degree of similarity in order to facilitate mapping consistency, semantic interpretation, learning, and memorization. In many applications of spatial or temporal visualization, there are quite often a large number of small glyphs required. Sometimes they are overlaid on some background graphics or imagery. Hence the *differentiability* of glyphs and potential perceptual errors in observation and exploration is a critical issue in glyph design.

Figure 2.18 shows example cases that may render some glyphs indistinguishable. Zooming-out actions in data exploration can reduce glyph size significantly. For example, they could make some shapes (e.g., circle and hexagon) and textures appear similar, while confusing the categorization of sizes (e.g., big, medium, small). Meanwhile, environmental lighting conditions and printing or photocopying facilities can cause color and greyscale degeneration. Not only would such changes make some glyphs indistinguishable, but would also confuse the association between different colors or grayscales. Whilst a dynamic legend may help alleviate the confusion about various mappings, it demands users to view the legend on a regular basis, incurring additional cognitive load in terms of the effort for the bothersome visual search and memorization of the unstable mapping keys. Other issues could also include color- or change-blindness, short- or long-sightedness, clustering, occlusion, distortion, and so on.

2.4.2.2 Hamming Distance

In data communication, a *code* consists of a finite set of *codewords*, each of which is a digital representation of a letter in an alphabet. In the context of binary encoding, *Hamming distance*, proposed by Richard Hamming in 1950

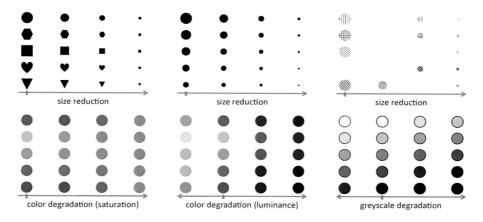

Figure 2.18 Three different types of quality degradation are applied to several glyphs, each of which is encoded using a single visual channel. The original quality is indicated by a marker on the x-axis. When size, saturation, and luminance are changed, they become more difficult to differentiate [79].

[59], is a measure of the number of bit positions in which two codewords differ. Considering all pairs of codewords in a code, the minimal distance is referred to as the minimal Hamming distance of the code. (In the literature, the word *minimal* is often confusingly omitted.) In communication, there are two main strategies for handling errors that occur during transmission. *Automated error detection* allows the receiver to discover that any error has occurred and to request a retransmission accordingly.

Automated error correction enables the receiver to detect an error and deduce what the intended transmission must have been. Hamming defined the following principle:

Theorem. A code of $d+1$ minimal Hamming distance can be used to detect d bits of errors during transmission. A code of $2d+1$ minimal Hamming distance can be used to correct d bits of errors during transmission [59].

For example, given a 3-bit code as illustrated in Figure 2.19, there are 8 possible codewords. One may select a subset of these codewords to construct a code with its minimal Hamming distance equal to 2 bits or 3 bits. Figure 2.19(a) shows one of these codes, which has 4 codewords and is of 2 bits Hamming distance. This code can detect 1-bit errors since any change of a valid codeword by 1 bit would result in an invalid codeword, which would lead the receiver to discover the error.

Figure 2.19(b) shows another code with 2 codewords, and is of 3 bits Hamming distance. It can detect 2-bit errors and correct 1-bit errors. When

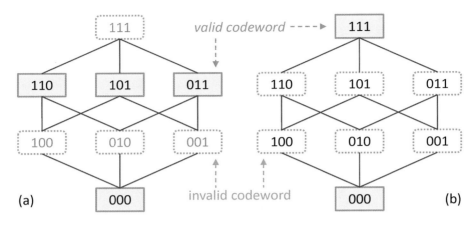

Figure 2.19 Two 3-bit codes. (a) A code can detect 1-bit errors. (b) A code can detect 2-bit errors and correct 1-bit errors.

a valid codeword (e.g., 111) is changed by 1 bit during transmission (e.g., 110), the receiver can detect such an error and recover the intended codeword based on the nearest-neighbor principle. Of course, if a 2-bit error occurred during transmission, the receiver would be able to detect the error but could not make a correct "correction." Nevertheless, if it is known that 2-bit errors are likely to occur, then this should either be used as only an error detection code, or a code with a longer Hamming distance should be used instead.

2.4.2.3   Quasi-Hamming Distance for Glyph Design

Similar to data communication, a set of glyphs is a code (i.e., alphabet), and each glyph in the set is a valid codeword (i.e., letter). During visualization, there can be errors in displaying or perceiving a glyph. If a viewer can detect that a perceived glyph is not quite "right," conscious or unconscious effort can be made to correct such an error. Conscious effort, which is an analogy of error detection and repeated transmission, may typically include zooming in to have a closer look, or consulting the legend. Unconscious effort, which is an analogy of error correction, may include some gestalt effects [26], and inference from other visual information [95].

Figure 2.20 shows two example glyph sets, each with 8 codewords. Given the two display errors depicted on the left, i.e., an arrow glyph is skewed in a distorted printout and a shape glyph is occluded by another shape, one can detect both errors easily. The error with the arrow glyph may need some conscious effort, and that with the shape error can usually be corrected unconsciously. This suggests that it is possible to establish a conceptual framework, similar to Hamming distance, for error detection and error correction in glyph-based visualization.

Figure 2.20 Two examples that illustrate the phenomena of error detection and error correction in glyph-based visualization. (Above) A viewer may sense that the glyph on the left may not be correct in a distorted visualization, and consult the legend to correct the error. (Below) A viewer may unconsciously perceive the glyph on the left as a star shape due to gestalt effects and a priori knowledge about the glyph set.

However, understandably, measuring the distances and errors in visual perception is not as simple as measuring those represented by binary codewords. We thereby propose an approximated conceptual framework based on the principle of Hamming distance, and we call it *quasi-Hamming distance* (QHD). The term *quasi* implies that the distance measure is approximated, and the quantitative measure of perceptual errors is also approximated. The main research questions are thereby (i) whether we can establish a measurement unit common to both measures, and (ii) how we can obtain such measurements.

The answer to the first research question is that we can utilize the "bit" as the common unit for both distance and error measurement. Let us first consider an ordered visual channel, such as brightness or length, as a code C. Theoretically, C can have a set of codewords $\{c_1, c_2, \ldots c_n\}$ such that the difference between two consecutive codewords is the just-noticeable difference (JND) of this visual channel. We can define the QHD between each pair of codewords c_i and c_j as $|i - j|$ bits. During display and visualization, if c_i is mistaken for c_j, we can call this a d-bit error where $d = |i - j|$. Now let us extend this concept to a less-ordered visual channel (e.g., hue) or an integrated channel (e.g., color). Theoretically, we can construct a code $\mathbb{C}$ by uniformly sampling the space of the visual channel (e.g., the CIE L*a*b* color space) while ensuring that every pair of samples differ by at least the JND of this channel. These codewords, i.e., samples, can be organized into a network, where the distance between any two codewords can be approximated proportionally according to the JND (i.e., JND = 1 bit). Note that the possible perception error rate with a code that maximizes the number of codewords based on the JND is likely to be very high. In practice, one designs a glyph set based on only a small subset of samples in a visual channel or more commonly in the multivariate space of several visual channels. Hence, a QHD measure based on JND would be too fine to use in practice, though in a longer term,

JND can provide an *absolute reference measure* once we have obtained such measures for most visual channels in visualization.

This leads to the second research question, i.e., given a glyph set, how can we measure the distance between glyphs? One may consider using the following methods:

1. **Estimation by expert designers.** This practice has always existed in designing exercises such as for traffic signs and icons in user interfaces. To formalize this practice, designers can explicitly estimate and label the distance between each pair of glyphs in a glyph set. While this approach may be most convenient to the designers, its effectiveness depends very much on the experience of the designers concerned and it is rather easy to overlook certain types of display and perception errors.

2. **Crush tests.** One can simulate different causes of errors, such as those illustrated in Figure 2.18, and determine at which level of degeneration glyphs may become indistinguishable. The corresponding level of degeneration can be defined as QHD. While this approach would yield more consistent estimation of QHD, more research would be required to compile a list of different causes of errors and define coherent levels of degeneration across different causal relationships.

3. **Task-based evaluation.** Similar to (2), one can simulate different visualization conditions, enlist users to perform their tasks, measure users' performance, and transform performance measures to QHD. On the one hand, this approach is perhaps most semantically meaningful for a particular glyph set in a specific application context. On the other hand, the performance measures collected may feature many confounding effects, while there may only be a small number of users available for such an evaluation.

4. **User-centric estimation.** One may conduct a survey among human participants about how easy or difficult it is to differentiate different glyphs. By removing task-dependency in (3), more participants can be involved in such a survey, yielding a more reliable estimation of QHD.

5. **Computer-based similarity measures.** There is a large collection of image similarity measures in the literature [101, 150]. In a longer term, it is likely that we will be able to find measures that are statistically close to user-centric estimation, though there is not yet a conclusive confirmation of optimal image similarity measures, and there are hardly any metrics specially designed for measuring similarity of glyphs.

The concept of QHD was applied to a practical application, where a glyph-based visualization tool was developed for observing activities in file systems [78]. Since the visualization tasks had to handle a high volume of data, the glyphs were expected to be relatively small and plentiful on a display screen.

In addition, the design of the glyphs features metaphoric encoding to assist in learning and memorization. It was necessary to ensure that the glyphs would be differentiable easily in routine observational visualization.

After producing the initial design of a set of glyphs, Legg et al. employed several ways to estimate the QHDs among the glyphs, including expert estimation, crush tests, a survey (user-centric estimation), and an image similarity metric (computer-based similarity measures) [78]. The estimated QHDs were used to improve the design. The glyph-based visualization software was further tested using real-world data, e.g., Dropbox Activity Log and Git Repository History.

2.4.3   User Studies in the Information-Theoretic Framework

In previous sections, we have shown that information theory provides a quantitative measure of information in a visualization context. We do not, however, suggest that such quantitative measurements might replace user studies. On the contrary, the above discussions have naturally led us to question the role of user studies from the perspective of information theory, since user studies produce statistics about phenomena and events in visualization.

In an information-theoretic framework, user studies have a much bigger role than in state-of-the-art of visualization. A fundamental component of any information-theoretic measure is the *probability mass function*. Perhaps we may estimate such a function for an input variable X based on our domain knowledge about the application concerned, or we may obtain this by placing a data flow monitor in the *vis-encoder* part of the visualization pipeline (Figure 2.2). However, we simply do not have sufficient knowledge about human perception and cognition to estimate such a function yet.

Recall the overview+detail example in Section 2.2, the two joint probability mass functions in Table 2.2 are synthetic data to demonstrate a mathematical concept. However, such data can be collected through user studies, and to a certain extent, may also be collected seamlessly through users' interaction with the system. The challenges will be our understanding of which probabilistic attributes are fundamental and generic in visualization, so we can estimate a finite set of probability mass functions to be used in practical applications of information theory. For example, in language processing, we have statistics about probability of the appearance of each English letter, the conditional probability about one letter after another, the redundancy in printed English, and so on. Such statistically estimated probability mass functions have been used effectively in applications such as data compression and hand-writing recognition. If we have such fundamental statistical findings from visualization user studies, we can transfer information theory to practice in visualization.

2.5 SUMMARY

Visualization is concerned with visually coding and communicating information. Many aspects of a visualization pipeline feature events of a probabilistic nature, bearing a striking resemblance to a communication pipeline. It is a reasonable assumption that the science of visualization should be built upon a number of theories established in different disciplines. It is also rational to consider information theory as one of these theories.

In this chapter, we showed that one can apply information-theoretic measures to several aspects of visualization. We showed that the major concepts of information theory, ranging from mutual information to multiplexing, and from data processing inequality to redundancy, are all relevant to visualization. We showed that information theory can explain many complex phenomena in visualization (e.g., logarithmic plots and visual multiplexing). We showed that information theory can be used to guide the optimization of visualization process (i.e., cost–benefit analysis). So far, we have not discovered any other mathematical theory that could offer such a broad underpinning framework.

FURTHER READING

Chen, M. and Floridi, L. (2013). An analysis of information in visualisation. *Synthese*, 190(16):3421–3438.

Chen, M. and Golan, A. (2016). What May Visualization Processes Optimize? to appear in *IEEE Transactions on Visualization and Computer Graphics*, doi:10.1109/TVCG.2015.2513410.

Chen, M. and Jänicke, H. (2010). An information-theoretic framework for visualization. *IEEE Transactions on Visualization and Computer Graphics*, 16(6):1206–1215.

Chen, M., Walton, S., Berger, K., Thiyagalingam, J., Duffy, B., Fang, H., Holloway, C., and Trefethen, A.E. (2014). Visual multiplexing. *Computer Graphics Forum*, 33(3):241–250.

CHAPTER 3

Viewpoint Metrics and Applications

CONTENTS

3.1   VIEWPOINT METRICS AND BASIC APPLICATIONS

Automatic selection of the most informative viewpoints from which a 3D model or a dataset is rendered is a very useful focusing mechanism in visualization, guiding the viewer to the most interesting information of the scene or dataset. A selection of the most informative viewpoints can be used for a virtual walkthrough or a compact representation of the information the data contains. The best view selection algorithms have been applied to computer graphics domains, such as scene understanding and virtual exploration [1, 3, 27, 44, 91, 105, 112, 132], N best views selection [44, 83, 133], volume visualization [8, 11, 20, 68, 99, 117, 136, 138], image-based modeling and rendering [47, 83, 132], flow visualization [119], shape retrieval [7, 43, 53, 80], mesh simplification [18, 19], molecular visualization [102, 131], and camera placement [91].

Information theory measures, such as entropy and mutual information, have been used in computer graphics and scientific visualization to measure the quality of a viewpoint. Viewpoint entropy, first introduced by Vázquez et al. [130] for polygonal models, has been applied to volume visualization by Bordoloi and Shen [8] and Takahashi et al. [117]. In particular, Bordoloi and Shen [8] obtained the goodness of a viewpoint from the entropy of the visibility of the volume voxels. Viola et al. [136] proposed a visibility channel and used the viewpoint mutual information to automatically determine the most expressive view on a selected focus. A unified information-theoretic framework for viewpoint selection, ambient occlusion, and mesh saliency for polygonal models has been presented in [44, 54]. Tao et al. introduced in [119] the application to flow visualization and exploration. Viewpoint measures for flow visualization will be presented in Chapter 5.

Approaches for the estimation of the most informative viewpoints for scientific datasets are similar to those developed for polygonal data, with differences due to the different focus of scientific visualization with respect to more general computer graphics. One main difference from polygonal computer graphics is that the underlying data is in general more complex, as scientific data are generated by measurements and simulations that have a very heterogeneous output. Some data contain solely measurements in numerical values without a priori knowledge about structures, while other data contain information about the most relevant structures such as critical points in flow data or segmentation masks of anatomical objects. The heterogeneity among scientific data types corresponds to the heterogeneity in the visualization approaches for viewpoint quality evaluation.

3.2 FROM POLYGONS TO VOLUMES

View selection algorithms were originally designed for polygonal data. Volumetric datasets, that allows the use of semi-transparent occlusion, consist in general of an order of magnitude more data elements as compared to a scene described by polygons. Thus, modifications to original techniques are necessary, both to take into account the magnitude of data and the visibility computation.

3.2.1 Isosurfaces

One of the first approaches in view selection for volume data was designed by Takahashi et al. [117] for fast evaluation of viewpoint quality based on the visibility of extracted isosurfaces or interval volumes. This approach represents a compromise between surface-based visibility estimation techniques, applied to polygons, and purely voxel-based visibility estimation.

The algorithm extracts isosurfaces by uniformly sampling the entire range of the scalar field values, and then takes the average of viewpoint entropies for the extracted isosurfaces. For each viewpoint v and each isosurface i ($i = 1...n$), A_{ij} denotes the visible area of the j-th face on the 2D screen as ($j =$

$1...m_i$), while the background area is denoted by A_{i0}. Since the total area S of the 2D screen remains constant for all the isosurfaces, we can formulate the (normalized) viewpoint entropy for the isosurface i as

$$E_i^{iso}(v) = \frac{-1}{\log(m_i + 1)} \sum_{j=0}^{m_i} \frac{A_{ij}}{S} \log \frac{A_{ij}}{S}. \tag{3.1}$$

Then, for each isosurface, locally optimal viewpoints can be computed using Equation 3.1.

The viewpoint entropy E^{iso} of the entire volume is computed as an average of the isosurface entropies, and is given by

$$E^{iso}(v) = \sum_{i=1}^{n} \frac{E_i^{iso}(v)}{n}, \tag{3.2}$$

which allows us to compute the optimal viewpoint as the one with the highest global viewpoint entropy. The viewpoint positioning is then improved as introducing the weighted sum of per-feature viewpoint entropies. This weighting by importance of feature components can be, for example, the average opacity of voxels in the interval volume specified by the transfer function $\overline{\alpha}_i$. Viewpoint quality considers occlusion among features in the calculation in order to avoid feature overlap in the case of the best viewpoints. While evaluating view quality of a specific feature, only the unoccluded regions of that interval volume are visible to the view selection calculation.

Thus, the weighted feature-driven viewpoint selection [117] is defined by

$$E_\lambda^{iso}(v) = \sum_{i=1}^{n} \frac{\lambda_i}{\sum_{j=1}^{n} \lambda_j} E_i^{iso}(v), \tag{3.3}$$

where the sum of viewpoint entropy $E_i^{iso}(v)$ for feature i and viewpoint v is weighted by an importance factor λ_i. This weighting factor can be, for instance, proportional to the average opacity $\overline{\alpha}_i$, which integrates the volumetric information into the viewpoint selection process.

Figure 3.1 shows best and worst views of a set of interval volumes extracted from the hydrogen dataset including the bounding sphere plot indicating areas of varying viewpoint quality.

3.2.2 Volumetric Data

Visibility estimation using interval volumes enables fast computation of view quality of volumetric datasets. To compute a precise per-voxel visibility, however, the entire volume has to be considered in the evaluation. Therefore, Bordoloi and Shen [8] introduced viewpoint entropy for volumetric data as an adaptation from the viewpoint entropy approach for polygonal data [130]. For the probability distribution function, the area visibility distribution of polygonal faces is replaced by the visual probability $q_i(v)$ of voxel i obtained from

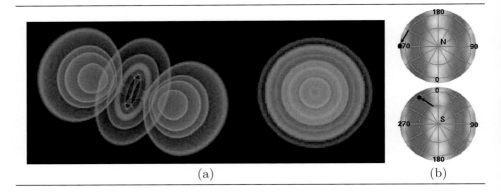

(a)   (b)

Figure 3.1   Viewpoint estimation for interval volumes [117]: (a) best and worst views of interval volumes extracted from a dataset containing simulated electron density distribution in a hydrogen atom; (b) viewpoint quality distribution map with a bright region encoding low–quality viewpoints and a dark region encoding the good and informative viewpoints. Worst (top) and best (bottom) views are indicated by black dots.

the fraction between the voxel visibility $\nu_i(v)$ for a given viewpoint v and the voxel importance W_i. The term visibility ν denotes the transparency of the material between the camera and the voxel and is equal to

$$\nu_i(v) = \frac{1}{N_{\mathrm{r}}} \sum_{r=1}^{N_{\mathrm{r}}} \prod_{k=1}^{N_{\mathrm{s}}(r)} (1 - tf(s_k).\alpha), \tag{3.4}$$

where N_{r} is the number of rays intersecting the voxel i; k is the sample iterator along the ray r from the first sample s_1, which is entering the volume, to the last sample $s_{N_{\mathrm{s}}(r)}$ before intersecting the voxel i; and $tf().\alpha$ is the opacity transfer function.

The importance distribution can, for example, be defined as the opacity value specified in the transfer function. This means that more opaque voxels will get more prominence than more transparent regions. A more advanced voxel relevance function can incorporate shape characteristics and color in addition to the opacity. The probability distribution function is defined as

$$q_i(v) = \frac{1}{\sigma} \frac{\nu_i(v)}{W_i}, \tag{3.5}$$

where σ is a normalization factor so that the probabilities add to 1,

$$\sigma = \sum_{i=1}^{N_{\mathrm{v}}} \frac{\nu_i(v)}{W_i}, \tag{3.6}$$

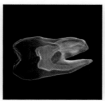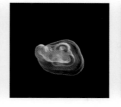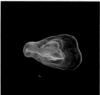

Figure 3.2 View selection for static volumes [8]: Two best views (left) and two worst views (right).

where N_v is the number of voxels.

Thus, the viewpoint entropy of viewpoint v is given by

$$H(v) = -\sum_{i=1}^{n} q_i(v) \log q_i(v). \tag{3.7}$$

Figure 3.2 shows the two highest entropy views (best views) and the two lowest entropy views (worst views) for the tooth dataset.

Having evaluated the viewpoint quality for a particular data type, volumetric or polygonal, it is natural that viewpoints nearby frequently have similar viewpoint qualities. When the visualization goal is to provide a set of representative viewpoints instead of a single best one, a viewpoint clustering scheme is needed, and information-theoretic measures for clustering views according to similarity can be used. The viewpoint similarity is computed using the Jensen–Shannon divergence (Equation 1.30), i.e., the similarity between two views v_1 and v_2 with distributions $q(v_1)$ and $q(v_2)$, respectively, is given by

$$JS(\frac{1}{2}, \frac{1}{2}; q(v_1), q(v_2)) = H(\frac{1}{2}q(v_1) + \frac{1}{2}q(v_2)) - H(q(v_1)) - H(q(v_2)), \tag{3.8}$$

which is bounded between 0 and 1.[1] According to the similarity values, views either belong to the same cluster, or in case of strong dissimilarity, they each belong to a separate cluster. For all clusters, the most representative viewpoint is selected. Figure 3.3 shows a tooth dataset from several viewpoints that together capture most of the information about the scene.

The method by Bordoloi and Shen [8] was improved by Ji and Shen [68] by considering the distribution of the screen pixels, rather than the voxels, and combining with the information conveyed by color and curvature properties. Specifically, for viewpoint v, the probability distribution due to opacity is defined, for each pixel i from n pixels on the image seen from v, as

$$p_i(v) = \frac{\alpha_i(v)}{\sum_{i=1}^{n} \alpha_i(v)}, \tag{3.9}$$

---

[1] Actually, the measure used in [8] was two times $JS(1/2, 1/2; q(v_1), q(v_2))$, bounded thus between 0 and 2.

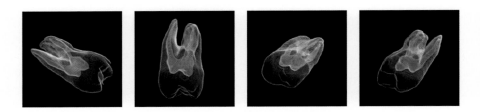

Figure 3.3 View selection for static volumes [8]: Four selected views of the tooth dataset from four bounding sphere partitions.

where $\alpha_i(v)$ is the opacity projected on pixel i.

Thus, the image entropy due to opacity can be defined as the entropy of distribution (Equation 3.9) and, after normalizing, we obtain the contribution of opacity, $opacity(v)$, given by

$$opacity(v) = \frac{1}{\log M} H(p(v)),$$ (3.10)

where M is the maximum projection size (in pixels) of the images among all the views.

Next, a color distribution is defined,

$$p_i(v) = \frac{A_i(v)}{\sum_{i=1}^{n} A_i(v)} = \frac{A_i(v)}{A_T},$$ (3.11)

where for each color we consider the screen area $A_i(v)$ covered by this color divided by the total screen area $A_T = \sum_i A_i$, and the entropy of this color distribution can be computed. After normalizing, the color entropy is given by

$$color(v) = \frac{1}{\log C} H(p(v)),$$ (3.12)

where C is the maximum number of colors. The contribution of the curvature, $curvature(v)$, is computed with the method of Kindlmann et al.[73]. In addition, voxels with higher curvature are assigned higher–intensity colors, which reflect on the resulting projected image. Finally, taking a convex combination of all three, $\beta_1 + \beta_2 + \beta_3 = 1$, $\beta_i \geq 0$, we obtain the quality of viewpoint v given by

$$u(v) = \beta_1 opacity(v) + \beta_2 color(v) + \beta_3 curvature(v).$$ (3.13)

In practice, the weights are chosen based on the characteristic of the data and the transfer function and the nature of the application. In Figure 3.4, the best and worst views are obtained considering only the opacity entropy, and in Figure 3.5, using the color entropy. In Figure 3.6, the best and worst views are computed using a combination of 80% curvature and 20% opacity.

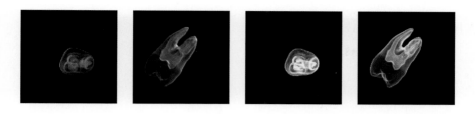

Figure 3.4 Tooth dataset with view selection using an opacity map [68]: From left to right, worst view, best view, worst view opacity map, best view opacity map.

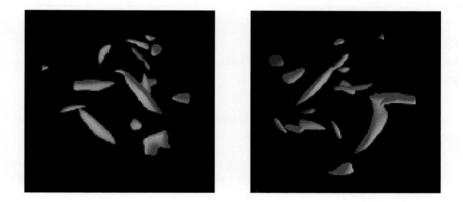

Figure 3.5 Vortex dataset with view selection using a color map [68]: Left, worst view, right, best view.

3.3 VISIBILITY CHANNEL IN VOLUME VISUALIZATION

Similar to the visibility channel for 3D polygonal models [44], an information channel can be defined between viewpoints and volume data.

3.3.1 Visibility Channel

To select the most relevant views of a volume dataset, a viewpoint quality measure, the viewpoint mutual information, can be defined [136] by considering an *information channel* $V \rightarrow Z$ between random variables V (the set of viewpoints $\mathbb{V}$) and Z (the set of objects or voxels $\mathbb{Z}$ of a volume dataset); see Figure 3.7(a). Viewpoints are indexed by v and voxels by z.

The information channel $V \rightarrow Z$ is characterized by a probability transition matrix (or conditional probability distribution) that determines, given

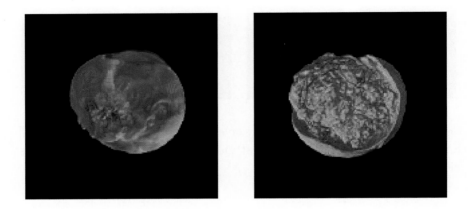

Figure 3.6 Terascale Supernova Initiative (TSI) dataset, with view selection using a combination of curvature and an opacity map [68]: Left, worst view, right, best view.

the input, the output probability distribution (see Figure 3.7(b)). The main elements of this channel are as follows [99, 136]:

1. The transition probability matrix $p(Z|V)$ is built from the conditional probabilities $p(z|v)$, with $\sum_z p(z|v) = 1$, computed as the quotients $\frac{\nu_z(v)}{\nu(v)}$, where $\nu_z(v)$ is the visibility of voxel z from viewpoint v, and $\nu(v) = \sum_z \nu_z(v)$ is the captured visibility of all voxels over the sphere of directions.

2. The input probability distribution $p(V)$ is given by the probabilities of selecting each viewpoint, where $p(v)$ can be interpreted as the impor-

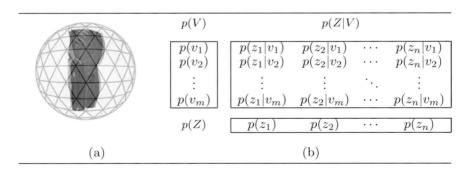

Figure 3.7 Visibility channel [99]. (a) Sphere of viewpoints of a voxel model. (b) Probability distributions of channel $V \to Z$.

tance of viewpoint v. Ruiz et al. [99] obtained $p(V)$ from the normalization of the captured visibility of the dataset over each viewpoint, while Viola et al. [136] assigned uniform importance to each viewpoint.

3. The output probability distribution $p(Z)$ is given by

$$p(z) = \sum_v p(v)p(z|v), \tag{3.14}$$

expressing the average visibility of each voxel.

The *mutual information* (MI) between V and Z, which expresses the degree of *dependence* or *correlation* between a set of viewpoints $\mathbb{V}$ and the volume dataset $\mathbb{Z}$, is given by

$$I(V; Z) = \sum_v p(V) \sum_z p(z|v) \log \frac{p(z|v)}{p(z)} = \sum_v p(v)I(v; Z), \tag{3.15}$$

where

$$I(v; Z) = \sum_z p(z|v) \log \frac{p(z|v)}{p(z)}, \tag{3.16}$$

the contribution of viewpoint v to mutual information $I(V; Z)$, is defined as the viewpoint mutual information (VMI), and measures the degree of dependence between viewpoint v and the set of voxels. The most *representative* viewpoint is defined as the one that has minimum VMI, as low values correspond to more independent views, showing the maximum possible number of voxels in a way that the visibility distribution $p(Z|v)$ of v is similar to $p(Z)$. This similarity is given by the Kullback–Leibler distance between $p(Z|v)$ and $p(Z)$ (Equation 1.12, see [44, 136]), which is zero when $p(Z|v) = p(Z)$. High values of $I(v; Z)$ mean a high dependence between viewpoint v and the object, indicating a highly coupled view, as between the viewpoint and a small number of voxels with low average visibility. One of the main properties of VMI is its robustness to deal with any type of discretisation or resolution of the volume dataset [136]. A similar behavior can be observed for polygonal data [44].

The viewpoint entropy (VE) of viewpoint v is defined by

$$H(Z|v) = -\sum_z p(z|v) \log p(z|v). \tag{3.17}$$

VE measures the degree of uniformity of the visibility distribution $p(Z|v)$ at viewpoint v. The best viewpoint is defined as the one with maximum VE, which would happen when a certain viewpoint can see all the voxels with the same projected visibility. Minimum VE would be obtained when most of the visibility is captured from a few voxels.

$$
\begin{array}{cc}
p(Z) & p(V|Z) \\
\begin{array}{|c|}\hline p(z_1) \\ p(z_2) \\ \vdots \\ p(z_n) \\ \hline\end{array} &
\begin{array}{|cccc|}\hline p(v_1|z_1) & p(v_2|z_1) & \cdots & p(v_m|z_1) \\ p(v_1|z_2) & p(v_2|z_2) & \cdots & p(v_m|z_2) \\ \vdots & \vdots & \ddots & \vdots \\ p(v_1|z_n) & p(v_2|z_n) & \cdots & p(v_m|z_n) \\ \hline\end{array} \\
p(V) & \begin{array}{|cccc|}\hline p(v_1) & p(v_2) & \cdots & p(v_m) \\ \hline\end{array}
\end{array}
\rightarrow
\begin{array}{c}
MIM(Z;V) \\
\begin{array}{|cccc|}\hline I(z_1;v_1) & I(z_1;v_2) & \cdots & I(z_1;v_m) \\ I(z_2;v_1) & I(z_2;v_2) & \cdots & I(z_2;v_m) \\ \vdots & \vdots & \ddots & \vdots \\ I(z_n;v_1) & I(z_n;v_2) & \cdots & I(z_n;v_m) \\ \hline\end{array}
\end{array}
$$

(a) (b)

Figure 3.8 (a) Probability distributions of channel $Z \rightarrow V$, used to compute the voxel mutual information. (b) The elements of matrix $MIM(Z;V)$ are given by $I(z_i;v_j) = p(v_j|z_i)\log\frac{p(v_j|z_i)}{p(v_j)}$ and used to calculate the color ambient occlusion in Section 3.6.1 [99].

3.3.2 Voxel Information

In Section 3.3.1, we obtained the information associated with each viewpoint (VMI) from the information channel $V \rightarrow Z$, between the sphere of viewpoints and the voxels (or objects) of the volume dataset. Ruiz et al. [99] introduced the voxel information from the reversed channel $Z \rightarrow V$, so that Z is now the input and V the output; see Figure 3.8(a). Note that the MI is invariant to the reversion of the channel: $I(V;Z) = I(Z;V)$. The idea of reversing the viewpoint channel appeared first in [54] for polygonal models, together with the computation of the information associated with a polygon.

From Bayes' theorem $p(v,z) = p(v)p(z|v) = p(z)p(v|z)$, the MI (Equation 3.15) can be rewritten as

$$I(Z;V) = \sum_z p(z) \sum_v p(v|z)\log\frac{p(v|z)}{p(v)} = \sum_z p(z) I(z;V), \tag{3.18}$$

where

$$I(z;V) = \sum_v p(v|z)\log\frac{p(v|z)}{p(v)} \tag{3.19}$$

is the contribution of voxel z to $I(Z;V)$ and is defined as the voxel mutual information (VOMI). This represents the degree of correlation between the voxel z and the set of viewpoints, and can be interpreted as the information associated with voxel z. As for the VMI, low values of VOMI correspond to voxels seen by a large number of viewpoints in a balanced way, that is, voxels with conditional probability distribution $p(V|z)$ similar to $p(V)$. The opposite happens for high values.

The mutual information matrix $MIM(Z;V)$ in Figure 3.8 is constituted by the terms $I(z;v) = p(v|z)\log\frac{p(v|z)}{p(v)}$, which appear in Equation 3.18, where each term represents the shared information between voxel z and viewpoint v.

Figure 3.9 shows, for different datasets, the VOMI maps computed using 42 viewpoints and colored using the thermal scale represented in Figure 3.9(e). Warm colors correspond to high VOMI values and cool colors to low ones.

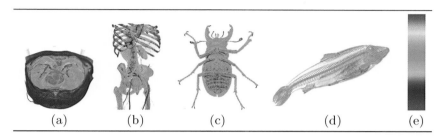

(a)　　　　(b)　　　　(c)　　　　　　(d)　　　　(e)

Figure 3.9　VOMI maps generated using 42 viewpoints for different models and transfer functions: (a) CT-body, (b) CT-body (skeleton), (c) CT-beetle, and (d) CT-salmon. Maps are colored using the thermal scale in (e) [99].

The VOMI measure allows varied interpretations that can be used in different visualization applications, such as viewpoint selection (Section 3.5), volume illustration (Section 3.6), and transfer function design (Section 4.5).

Figure 3.10 shows the steps to compute the VOMI of a voxel model [99]. After the volume data is classified using a transfer function, a ray casting is performed, with the volume dataset centered in a sphere of viewpoints and the camera looking at the center of this sphere. In this way, a histogram of visibilities is created for each viewpoint, allowing us to estimate $p(Z|v)$. Using Equation 3.14 and Bayes' theorem, $p(Z)$ and $p(V|Z)$ are obtained from $p(V)$ and $p(Z|V)$. Finally, the VOMI map is obtained.

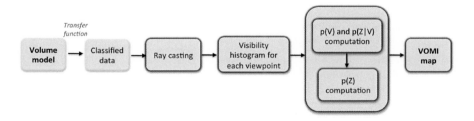

Figure 3.10　Overview of the VOMI pipeline [99].

3.4　IMPORTANCE-DRIVEN FOCUS OF ATTENTION

In volume visualization, a region of interest (ROI) is defined by interpretation and segmentation operations. Voxels are grouped into clusters forming domain

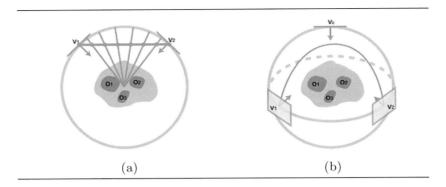

(a)   (b)

Figure 3.11 Viewpoint navigation with focus of attention [136]: (a) The viewpoint path is calculated as a difference between two viewpoint positions and normalized onto the bounding sphere, with smooth acceleration and deceleration in viewpoint change. (b) Approximating the trajectory between two optimal viewpoints by a Bezier curve using a contextual viewpoint v_c.

objects. These objects are often of high interest to users and good viewpoints are generally those that clearly depict object's structure. The visualization task can be, for example, to communicate the data interpretation to another audience. For such a task, viewpoint selection can be utilized to implement a high-level interaction request such as: "Show me object X." View selection therefore should not be based solely on the visibility of graphic primitives (e.g., voxels). To provide the best view of a certain feature, the visibility of these objects has to be computed. The simplest and fastest way to compute the optimal viewpoint of domain objects can be along the lines of the feature-driven viewpoint estimation approach [117].

Importance-driven focus of attention [136] is one example where guided navigation through pre-classified features in a volumetric dataset uses automatic viewpoint selection for setting the view to the most informative viewpoints on a given object of interest. The object in focus is directly selected by the user by choosing it from a list of classified objects. A characteristic viewpoint for this object is selected in combination with a visually pleasing discrimination of the focus from the remaining context information. By changing the object of interest, both viewpoint settings and visual parameters are smoothly modified to put emphasis on the newly selected object of interest. Figure 3.11 shows the change in viewpoint. The most informative viewpoints of each structure are pre-computed using viewpoint mutual information (Equation 3.16). This measure was used instead of the more common viewpoint entropy as it has better properties when dealing with view selection for importance-weighted objects within the volumetric data. For a specific viewpoint position v and set of objects $\mathbb{O}$, the VMI expresses the level of correlation of object visibility with respect to the viewpoint position. The best views of an object are those with the lowest mutual information. This is in contrast to the view evaluation using viewpoint entropy, which is maximal for the best views. The low VMI expresses that the viewpoint change has a small influence on the object visibility and thus the viewpoint is stable. Observe that VMI, Equation 3.16, can be written as a Kullback–Leibler distance (we consider now objects O instead of voxels Z)

$$I(v; O) = D_{KL}(p(O|v)||p(O)), \qquad (3.20)$$

where $p(O|v)$ is the conditional probability distribution between v and the dataset, and $p(O)$ is the marginal probability distribution of O. Thus, $I(v; O)$ is the relative entropy or Kullback–Leibler distance (Equation 1.12), between the visibility distribution of objects from viewpoint v and their average visibility. The smaller the measure the better the viewpoint, as we approach the ideal target of viewing every object as proportional to the average visibility $p(o)$. In this case, $I(v; O)$ would be zero. Adding importance to our scheme means simply modifying the target function. The ideal viewpoint would now be the one viewing every object as proportional to the average visibility times importance. Thus, viewpoint mutual information $I'(v; O)$ for importance-weighted data objects is defined as

$$I'(v; O) = D_{KL}(p(O|v)||p'(O)) = \sum_{o \in \mathbb{O}} p(o|v) \log \frac{p(o|v)}{p'(o)}, \qquad (3.21)$$

where

$$p'(o) = \frac{p(o)imp(o)}{\sum_{o \in \mathbb{O}} p(o)imp(o)},$$

and $p(o|v)$ is the data object o visibility from viewpoint position v and $p(o)$ is the marginal probability (i.e., the sum of object visibilities from all viewpoints and $imp(o)$ is the importance of object o). The importance allows us to flexibly integrate focusing on object(s) of interest in the view selection computation. One, two, or more objects can be assigned a high importance value. All these objects will then have high visibility in the selected viewpoint. See [44] for importance in viewpoint for polygonal data.

Finding a viewpoint where the characteristics of a specific feature are clearly visible requires a visibility estimation scheme. Complete visibility information requires ray casting of the whole dataset from various viewpoints, similar to the view selection for volumes based on viewpoint entropy computed from voxels [8]. The visibility computation is based on the opacity contribution of each voxel and object visibility is computed as the sum of voxel visibilities corresponding to the object. Additionally, two weights influence the visibility of an object (i.e., image-space weight and object-space weight). Image-space

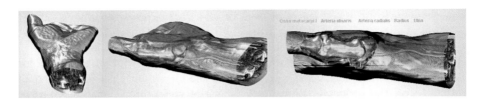

Figure 3.12   Importance-driven focusing [136]: Smooth guided navigation to focus on the object of interest.

weight penalizes the visibility of objects when they are located outside the image center. Object-space weight assigns higher visibility to objects that are closer to the viewing plane and penalizes those that are farther away.

The final object visibility is then mapped to a conditional probability of the object for a given viewpoint $p(o|v)$. These values are used for the computation of good viewpoints for a given object by using the viewpoint mutual information weighted with object importance information.

After selecting visual representations of objects and identifying representative viewpoints, the crucial information to perform interactive focus of attention is available. During the guided navigation, the object in focus is assigned a higher importance value. This value is directly mapped to all focusing mechanisms: dense visual style, level of ghosting of cut-away views, and position of the viewpoint. Thus, the viewpoint transformation is also controlled by the importance distribution, and its variation during guided navigation smoothly changes to the most informative viewpoint of the object in focus.

Importance-driven focus of attention guides the user's focus to the object of interest, while still permitting interactive viewpoint manipulation. Focusing on a specific feature in the human hand dataset is shown in Figure 3.12. The viewpoint smoothly changes from the most informative viewpoint for the entire volume to the viewpoint emphasizing the object of interest. As shown, parallel to the viewpoint change, the focus is discriminated from neighboring structures by using a different visual style and ghosting is employed to suppress occluding structures.

Guidance among diagnostically relevant viewpoints for intervention planning in various medical scenarios [86] is an example of how the basic technology of view selection is applied to address a particular problem in the medical imaging pipeline. The data is first interpreted into anatomical objects and represented as polygonal meshes. The visibility of objects is calculated from these extracted isosurfaces. Good viewpoints are estimated using many parameters with adjustable influence: object entropy, importance of occluders, size of unoccluded surface, preferred view region by surgeons, distance to viewpoint, and viewpoint stability. The choice of parameters demonstrates as how tightly viewpoint estimation is bound to specific domains: The distance to an important feature defines the importance of other features (e.g., neck muscles

Figure 3.13 Semantics-driven view selection [86]: Guided navigation through features in the human neck dataset assists studying the correspondence between focus objects (i.e., lymph nodes and surrounding tissue such as neck muscle).

close to a lymph node that is in focus). Furthermore, guided navigation supports zooming to the object of interest. Guided navigation of lymph nodes for neck intervention planning is shown in Figure 3.13.

3.5 VIEWPOINT SELECTION USING VOXEL INFORMATION

The voxel information map, defined in Section 3.3.2, can be projected on a viewpoint v, weighting the VOMI of voxel z with the transition probability $p(v|z)$ and summing over all voxels. Using the values from Equation 3.19, the informativeness (INF) of a viewpoint v is defined by

$$INF(v) = \sum_z p(v|z)I(z;V),$$ (3.22)

representing the total amount of voxel information seen by each viewpoint. High values of INF correspond to viewpoints that see a lot of voxel information, i.e., highly occluded parts of the model. The regions with high voxel information values would show relevant details of the model. Low values of INF correspond to low voxel information, associated with smooth changes in visibility and with less detail.

As we have seen in Section 3.3, different information-theoretic viewpoint measures have been introduced to select the "best" views. However, how good a view is depends of our declared objective. If it is to see the maximum number of voxels, then viewpoint entropy (Equation 3.17) can be the most convenient measure. This is because the maximum entropy view would be obtained when all the voxels were seen with the same projected visibility. On the other hand, minimum entropy would be obtained when only one voxel was visible. Viewpoint mutual information (Equation 3.16) can be used to detect the most representative views, or the ones that are most similar to the virtual view of the object obtained from the projection of all viewpoints. The main difference between VE and VMI is that, while VE is very sensitive to the resolution of the dataset, VMI is robust to deal with any kind of segmentation [136].

Figure 3.14 shows the views that capture the maximum and minimum VE, VMI, and INF. For each model, the first row corresponds to the "best" views (maximum VE, minimum VMI, and maximum) and the second row to the "worst" views (minimum VE, maximum VMI, and minimum INF). Due to the regular discretization of the volume dataset in voxels, the behavior of VE and VMI is not significantly different, showing ,respectively , the maximum number of voxels in a uniform way and the most representative view. On the other hand, maximum INF aims to show the maximum number of highly occluded voxels.

3.6 APPLICATION TO ILLUSTRATIVE RENDERING

We consider here the application of voxel mutual information to volume illustration, by interpreting VOMI as an ambient occlusion and obtaining illustrative effects.

3.6.1 Ambient Occlusion

VOMI can be interpreted as an ambient occlusion (AO) term [54]. AO is a measure of the visibility around a voxel, but while classical AO takes into account only local visibility, VOMI measures the whole volumetric data visibility around a voxel, and how this visibility is distributed between the viewpoints surrounding the volume. More uniform visibility means that the viewpoint we consider is less important, meaning that the voxel is less interesting or informative, and the VOMI value low. On the other hand, less uniform visibility means that the viewpoint we consider is more important, meaning that the voxel is more interesting or informative, and has a high VOMI value.

To obtain the AO of each voxel, the VOMI of all voxels is normalized between 0 and 1 and subtracted from 1, as low values of VOMI, represented in the grey map by values near 1, correspond to non-occluded or visible (from many viewpoints) voxels, while high values of VOMI, represented in the map by values near 0, correspond to highly occluded voxels.

Figure 3.15 shows the AO maps corresponding to the models of column (a) generated using different strategies, where in columns (b) to (e), we can see the approaches by Landis [75], Stewart [115], obscurances [97], and VOMI [99]. Landis's approach is obtained by the application of the ambient occlusion technique [75] to volume rendering. The binary technique by Landis produces too contrasted an effect, due to sharp transitions within the discrete set of occlusion values, while the Stewart and obscurance methods generate smoother maps due to the continuous range of the occlusion values. These AO techniques take into consideration only the local occlusion of the voxel. The VOMI technique considers the whole visibility, and thus occlusions, from the voxel to all viewpoints, integrating information of the whole volume with respect to the given voxel. This information will result in an AO map that captures different effects from the volumetric model.

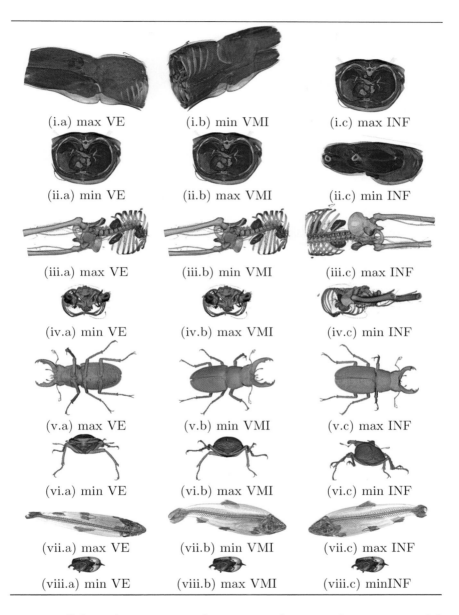

Figure 3.14 Selected viewpoints, from a set of 162, with various models according to (a) viewpoint entropy, (b) viewpoint mutual information, and (c) informativeness [99].

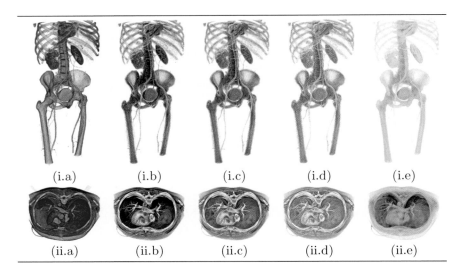

(i.a)	(i.b)	(i.c)	(i.d)	(i.e)
(ii.a)	(ii.b)	(ii.c)	(ii.d)	(ii.e)

Figure 3.15 AO maps generated using (b) Landis, (c) Stewart, (d) obscurances, and (e) VOMI approaches for the CT-body model with (a) two different transfer functions [99].

A first effect is obtained considering the AO value as an ambient lighting AL term. In this case, the color of a voxel z is obtained as

$$C(z) = AL(z) = k_i \, AO(z) \, C_{\mathrm{TF}}(z), \tag{3.23}$$

k_i being a constant factor that modulates the intensity of $AO(z)$, and $C_{\mathrm{TF}}(z)$ is the pure color of the voxel as defined in the transfer function.

Figure 3.16 illustrates the applications of the AO maps as an ambient lighting term, comparing the result of applying a local, classic ambient occlusion method [115] (Figure 3.16(b)) with the VOMI (Figure 3.16(c)). VOMI takes into account the whole volume visibility, offering a more shaded result than local ambient occlusion. This is clearly visible in the skeleton. The overall information given by the VOMI (Figure 3.16(c)) produces better results than local ambient occlusion (Figure 3.16(b)) with respect to the raw color information (Figure 3.16(a)).

Another effect is obtained by adding the AO term to the local lighting equation, as in the global illumination case where ambient occlusion mimics indirect illumination [62]. The color of a voxel is obtained as

$$C(z) = (1 - w_{\mathrm{AO}})((k_d N(z) \cdot L)C_{\mathrm{TF}}(z) + k_s(N(z) \cdot H)^n) + w_{\mathrm{AO}} AL(z), \tag{3.24}$$

where k_d and k_s are the diffuse and specular lighting coefficients, $N(z)$ is the normal of the voxel, L is the light vector, H is the half-angle vector between L and the direction to the viewer, $AL(z)$ is the ambient lighting, and w_{AO} is the weight of the ambient occlusion in the final color.

Figure 3.17 illustrates the application of the AO maps as an additive term to the local lighting, comparing the result of applying Stewart's method (Figure 3.17(b)) with the VOMI (Figure 3.17(c)). As in Figure 3.16, the VOMI (Figure 3.17(c)) produces better results with respect to the direct illumination image of column 3.17(a) than classic ambient occlusion (Figure 3.17(b)). The overall features of the volume model are more distinguishable, and context information is better captured, giving an enhanced depth perception, clearly visible in the ribs.

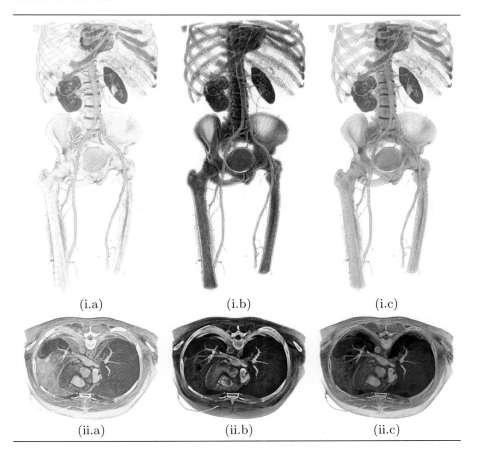

(i.a)   (i.b)   (i.c)

(ii.a)   (ii.b)   (ii.c)

Figure 3.16 The original CT-body model is shown (a) without illumination effects, and illuminated using AO computed with (b) Stewart's method and (c) VOMI, both applied as an ambient lighting term [99].

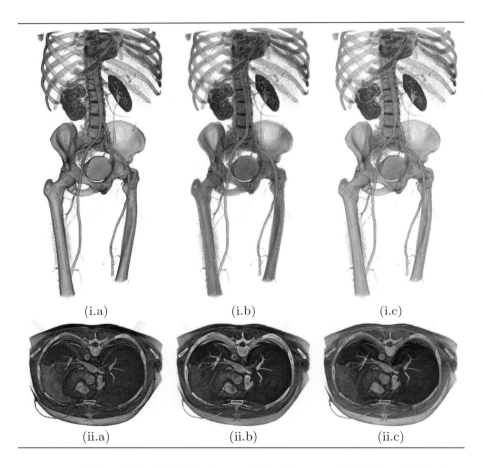

(i.a)   (i.b)   (i.c)

(ii.a)   (ii.b)   (ii.c)

Figure 3.17 The original CT-body model is shown with (a) local lighting, and with the AO computed with (b) Stewart's method and (c) the VOMI, both applied as an additive term [99].

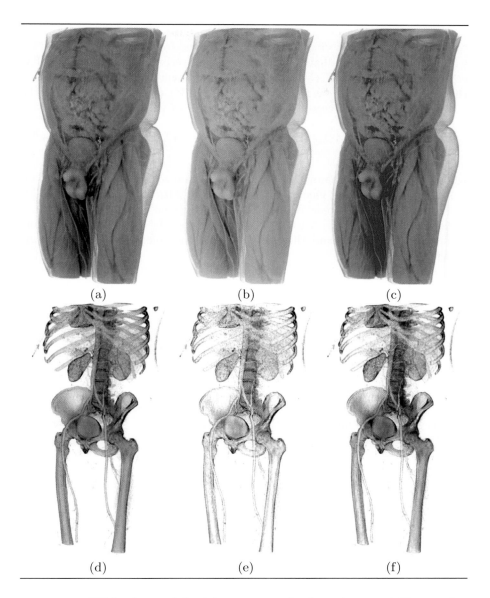

Figure 3.18 CT-body model with two transfer functions and illustrative effects: (a) grayscale AO map, (b) color AO map, (c) cool-and-warm AO map, (d) color AO map with contours, (e) and (f) different color AO maps with contours and color quantization [99].

3.6.2   Color Ambient Occlusion

The color ambient occlusion effect, which simulates the use of colored light sources at the different viewpoints, can be derived from the voxel information where, when all sources have the same color, we recover the original AO. The color ambient occlusion $CAO_\alpha(z; V)$ associated with voxel z is defined by the scalar product of row z of matrix $MIM(Z; V)$ (see Figure 3.8) and the complement of a color vector assigned to the set of viewpoints:

$$CAO_\alpha(z; V) = \sum_{v \in V} I(z; v)(1 - C_\alpha(v)), \qquad (3.25)$$

where α is the color channel, $C_\alpha(v)$ is the normalized color vector for channel α, and $I(z; v)$ is an element of $MIM(Z; V)$. After computing the VOMI for each channel, the color ambient occlusion is given by the combination of the color channel values. A color vector is obtained by assigning given colors to specific viewpoints and then interpolating the colors for the rest of the viewpoints.

The relighting effects can be combined with other illustrative effects, such as color quantization, contours, and cool-and-warm effects. Figure 3.18 shows some of these effects applied to the CT-body model, with two different transfer functions. Figures 3.18(a–c) show, respectively, the AO map, the corresponding color ambient occlusion, and the AO map colored using a cool-and-warm technique [55]. Figures 3.18(d–f) show the use of color ambient occlusion combined with contours and color quantization.

3.7   SUMMARY

In this chapter, we have shown the information-theoretic tools for viewpoint selection, and how the introduction of the concept of the information channel to model viewpoint selection allows us to obtain a unified framework from which to derive new applications.

In general, an automatic technique for obtaining good views on complex data is a useful tool and should be an integrated part of an entire visualization toolset. If we want a general applicability of this tool, the view selection approach should not be tailored to a very specific data or application scenario. The use of view selection based on information-theoretic measures, and in particular the use of the viewpoint information channel, allows us to cover all possible scenarios where visualization could be employed.

FURTHER READING

Bordoloi, U.D. and Shen, H.-W. (2005). View selection for volume rendering. In *IEEE Visualization 2005*, pages 487–494, IEEE, Minneapolis.

Ji, G. and Shen, H.-W. (2006). Dynamic view selection for time-varying volumes. *Transactions on Visualization and Computer Graphics*, 12(5):1109–1116.

Mühler, K., Neugebauer, M., Tietjen, C. and Preim, B. (2007). Viewpoint selection for intervention planning. In *Proceedings of Eurographics/ IEEE-VGTC Symposium on Visualization*, 267–274, Eurographics, Leeds.

Ruiz, M., Boada, I., Feixas, M. and Sbert, M. (2010). Viewpoint information channel for illustrative volume rendering. *Computers & Graphics*, 34(4):351–360.

Takahashi, S., Fujishiro, I., Takeshima, Y. and Nishita, T. (2005). A feature-driven approach to locating optimal viewpoints for volume visualization. In *IEEE Visualization 2005*, 495–502, IEEE, Minneapolis.

Viola, I., Feixas, M., Sbert, M. and Gröller, M.E. (2006). Importance-driven focus of attention. *IEEE Transactions on Visualization and Computer Graphics*, 12(5):933–940.

Volume Visualization

CONTENTS

It was predominantly the geometric mesh representation or image slices obtained from medical imaging modalities on which information-theoretic measures have found their first applicability in the domain of computer graphics and visualization. Soon after researchers investigated the usefulness of information-theoretic measures in the context of volume visualization. The basic data structure for volume visualization is the data volume, where at various positions in the 3D space, particular characteristics are encoded. The most elementary volumetric representation is an extension of the notion of a grayscale image that is defined on a two-dimensional grid, into 3D, forming a three-dimensional grid. On each grid location, a volume element (voxel) is storing information about the volumetric characteristics of the imaged or simulated phenomenon. Besides this so-called regular volumetric dataset, more complex variants of the volumetric representation may vary in data values over time, or can have multiple values stored per one volume element, or can be organized in an unstructured form instead of a regular-grid alignment.

Volume visualization utilizes various parameters that define the visual representation, such as the color and opacity values, defined in a so-called transfer function. This instrument assigns the optical properties to the values stored in each voxel enabling the user to visually inspect the most interesting data value intervals. As the volumetric datasets are of a dense nature, clipping and other smart visibility occlusion management techniques are necessary to reveal the structure of internal volume areas. To avoid tedious camera placement in three dimensions during an interactive 3D visualization, automated or guided viewpoint placement, as described in the previous chapter, simplifies the user's explorative tasks. Due to the nature of volumetric data being rather large, the effective design of a level-of-detail scheme or reduction of volumetric dataset into a set of representative isosurfaces are often ways to enable interactive volume visualization. This essential volume visualization toolbox can be automated by utilizing the information-theoretic measures. The following subsections look at individual volume visualization tools in detail.

4.1 TIME-VARYING DATA

In the previous chapter, view selection techniques for volumetric data were presented together with other data representations. As discussed, the classical viewpoint entropy [130] has been adapted to direct volume visualization by mapping the visibility, frequency, and opacity of a voxel to its so-called *visual probability* value as proposed by Bordoloi and Shen [8]. In this section, this topic is revisited in the context of view selection for time-varying data. For a single time step, the measure of entropy is used for estimating the best viewpoint on the volumetric data. For a time-varying dataset, the joint entropy is selected as a suitable measure, where each random variable from the joint set corresponds to the voxel visual probability distribution of a particular time step. The presented approach assumes a Markov sequence model, i.e., the information associated with time step t_i is only dependent on previous time steps t_{i-1}. Therefore the joint entropy (see Sections 1.1 and 1.4) can be written as

$$
\begin{aligned}
H(X) &= H(X_{t_1}, X_{t_2}, \ldots, X_{t_n}) & (4.1) \\
&= H(X_{t_1}) + H(X_{t_2}|X_{t_1}) + \ldots + H(X_{t_n}|X_{t_{n-1}}). & (4.2)
\end{aligned}
$$

The previously introduced importance factor (Section 3.2.2), defined by opacity and frequency of occurrence, is now extended to a conditional importance:

$$
W_j(t_i|t_{i-1}) = (k|\alpha_j(t_i) - \alpha_j(t_{i-1})| + (1 - k)\alpha_j(t_i)) \cdot I_j(t_i), \quad (4.3)
$$

where k is a parameter that assigns weight on the effect of the opacity change, $\alpha_j(t_i)$ is the opacity value of voxel j at the time step t_i. The information carried by the jth voxel is calculated as $I_j = -\log f_j$ and f_j is jth voxel probability or bin frequency. The conditional visual probability q_j of a voxel j at time step t_i given t_{i-1} and given a particular viewpoint V is calculated as

$$
q_j(V, t_i|t_{i-1}) = \frac{1}{\sigma} \cdot \frac{v_j(V)}{W_j(t_i|t_{i-1})}, \quad (4.4)
$$

where the σ is a normalization factor, $v_j(V)$ is the visibility of the voxel j from the viewpoint V and $W_j(t_i|t_{i-1})$ is the conditional importance factor defined above. This visual probability is then used for computing all the viewpoint conditional entropies for all but the first time step. The first time step is computed by the viewpoint entropy measure for volume data as described in

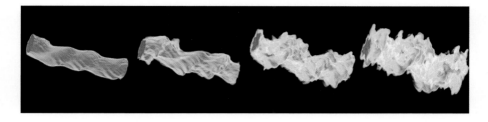

Figure 4.1 Optimal viewpoint estimated with the view selection approach for time-varying volume data based on the sum of conditional viewpoint entropy measures [8].

Section 3.2.2 of the previous chapter. The viewpoint position with the highest joint entropy is then selected as the viewpoint from which the temporal sequence can be best observed. The result of the highest view entropy measure for time-varying simulation data is shown in Figure 4.1. The shockwaves exhibit increasingly stronger turbulent flow over the simulation time, which can be clearly understood from the selected viewpoint.

For narrative visualization, an optimal camera *path* can sometimes be more interesting to determine than a single view *point*. Camera path planning is even more relevant in the context of time-varying volume visualization as the most interesting characteristics of the volume can change their position over time. The work by Ji and Shen [68] builds on the viewpoint selection techniques and presents an image-space view selection approach, in contrast to the object-space approach described above. Their approach is also described in detail in the previous chapter for static viewpoints. The following description is again related to time-varying volume data and the dynamic view selection for automatic camera path planning.

The information-theoretic measure that is employed for the best view selection is, in this case, the entropy measure that is computed for each view from the occurrence of the pixel opacities $H_{t_i,\alpha}(V_j)$, but also color values $H_{t_i,c}(V_j)$, and curvature information $H_{t_i,\kappa}(V_j)$. The final viewpoint quality of a viewpoint V_j at a time step t_i is a weighted sum of these entropy functions, denoted as the final utility function $u_{t_i}(V_j) = w_1 H_{t_i,\alpha}(V_j) + w_2 H_{t_i,c}(V_j) + w_3 H_{t_i,\kappa}(V_j)$, where the weights $w_1 + w_2 + w_3 = 1$. The goal of dynamic view selection is to find the viewing path that conveys the maximal amount of information about the temporal behavior within the dataset. Additional requirements are (1) the speed of the camera transformation should be nearly constant and (2) the direction of the camera viewing direction and position should not change abruptly. Finding a dynamic view that is optimal is a search problem of exponential complexity. However, given the above stated constraints related to the speed and direction, many implausible paths can be pruned and the search can be efficiently kept within bounds of polynomial time computation by a dynamic programming approach. The algorithm first calculates the information about the viewpoint quality $u_{t_n}(V_j)$ at the last time step t_n of the animation for each viewpoint $V_j | j \in [1..m]$ where m is the number of all viewpoints on a bounding sphere. This information is stored in the last row of a two-dimensional array called $u_{max}[t_i, V_j] | i \in [1..n] \times j \in [1..m]$ for all time steps t_i and all viewpoint positions V_j. Then, in the second step, all views for the time step t_{n-1} are evaluated. The $u_{max}[t_{n-1}, V_j]$ stores the sum of the most valuable path segment (computed from the last time step and the one before last time step), where the camera is at time t_{n-1} located at viewpoint V_j and in the following time step t_n, the viewpoint is in the neighborhood of V_j so that constant speed of the camera movement can be achieved. When the row t_{n-1} is filled with maximally valuable path segments, the analogous calculation is performed for the row t_{n-2} and so on until the row t_1. At the first time step, the maximal summed value along a particular path is selected as the best camera path. The calculation of the most valuable path segment is as follows:

$$u_{max}[t_i, V_j] = \max_{k \in [1..m]} u_{t_i}(V_j) - \phi(V_j, V_k) + u_{max}[t_{i+1}, V_k], \qquad (4.5)$$

where the $\phi()$ function evaluates whether the two viewpoint positions are close enough so that the camera speed can be maintained. If they are close enough $\phi() = 0$, otherwise $\phi() = \infty$. The $u_{max}[t_{i+1}, V_j] | j \in [1..m]$ row was calculated in the previous iteration and u_{t_i} is the final utility function.

However, this matrix only stores the camera path scores, not the connectivity between consecutive viewpoint positions along the camera path. This information is stored in another two-dimensional array so that when the corresponding highest value in $u_{max}[t_1, V_j]$ is identified for viewpoint V_j, in the same position of the related two-dimensional array, a link to the second viewpoint of the corresponding path at the time t_2 is stored. This way, the entire most valuable camera path can be reconstructed. However, there is still one issue to take into consideration. This algorithm allows for drastic turns along the camera path, which is clearly an undesired behavior. The above described algorithm can be extended by one more dimension, recording maximal values for each direction. This extension then allows us to finally select a camera path, where only mild changes to the direction from one viewpoint to another are possible.

The resulting camera path, which shows the most structural detail over an exemplary simulation sequence, is shown in Figure 4.2. The first image shows the camera trajectory and the series of rendered images are the most informative viewpoints are shown as key frames. The images show a high level of structural detail on the selected isosurface.

In the object-space view selection approach by Bordorloi and Shen [8], the data is characterized by its importance, which is essentially driven by its opacity value. The image-space approach has added color and curvature that together with the pixel opacity characterize an interesting feature. In the following, another technique is described that takes an object-space approach to

Figure 4.2 Dynamic view selection computed for the supernova simulation sequence shows that the viewpoint is smoothly changing so that the animation captures the most of evolving structural detail [68].

define interesting parts of time-varying data and uses a data value, and the local derivative properties. In contrast to the above view selection techniques, the goal of the technique introduced by Wang et al. [139] is to visually emphasize regions that are more interesting or change a linear playback of an animation so that more time is dedicated to interesting animation frames and the playback speed for frames that are not very informative is increased.

To achieve this, the time-varying volume data is first decomposed into spatial blocks that are rated based on interest. A block of data is considered as important or interesting if its content varies. If the content is characterized by a uniform value overall, then the block is not considered interesting. The importance within a block is assessed via the entropy measure (Equation 1.1). The random variable is the multidimensional feature vector that is formed by the data value, the gradient magnitude and direction, and the domain-specific derivatives. The probability of the random variable is captured in a normalized multidimensional histogram. Such characterization can be denoted as an *absolute* block importance. The higher the entropy value $H(X)$ is, the more interesting the data block is.

Furthermore, the importance is expressed in terms of how similar the block is to its neighboring blocks. If the blocks are largely similar to each other, their importance is not increased, but if a block is significantly different from its neighbors, then its importance, denoted as a *relative* block importance, increases. The relative block importance of block X, with a neighborhood around X denoted as S, is characterized by the similarity measure known as mutual information (Equation 1.10) $I(X; Y_i)$ such that $Y_i \in S$. The neighborhood can be defined as a spatial neighborhood S, meaning the blocks around block X, or it can be defined as a temporal neighborhood T with data blocks at the same spatial location as X, but at neighboring time steps. For this purpose, a normalized joint histogram data structure is created, which stores the joint probability of multidimensional features from both data blocks that are used for the mutual information computation.

After having the entropy computed as an absolute characterization of importance, and mutual information as a relative characterization of the importance with respect to the neighborhood, these two characterizations can be fused into one importance definition. A block is interesting if it has a high entropy value and low mutual information, because there is strong variation within the block and in comparison with the neighboring blocks. A block is uninteresting if it has low entropy and high mutual information, because the content is nearly constant and very similar to the neighborhood. Expressing the fused importance measure as a difference between $H(X)$ and $I(X;Y_i)$, the fused importance measure corresponds to the conditional entropy measure $H(X|Y_i)$. Taking into account the entire neighborhood, the importance value is calculated as

$$A(X, t_i) = H(X) - \sum_{Y_i \in S(or\ T)} w_i I(X; Y_i) = \sum_{Y_i \in S(or\ T)} w_i H(X|Y_i), \quad (4.6)$$

where w_i are the weights of the neighboring blocks with the constraint that $\sum_i w_i = 1$. An importance of a time step for all blocks is defined as a simple sum over all data blocks $A(t_i) = \sum_{X_j \in \mathbb{X}} A(X_j, t_i)$, where $\mathbb{X}$ is the set of all blocks.

When connecting the importance of a block over all time steps $A(X, t_i)|i \in [1..n]$ into an importance curve and plotting all importance curves in one time-importance plot, one can see the time-varying phenomena in a very compact, static form. It becomes more apparent how the importance values change over time. Having many data blocks, this depiction of lines, however, quickly results in visual clutter and it is hard to visually separate various temporal patterns of individual blocks. Clustering seems to be a way to address this problem and to understand classes of time-importance curves among the data blocks. The clustering approach, such as the k-means, can cluster the data blocks into classes of flow characteristics represented by the centroid curves. Taking one entire importance curve as the basis for clustering would however mean that flow characteristics for one block have to be identical to another block to form a cluster. This is a limiting constraint, as the flow through a block might exhibit different characteristics over the entire simulation sequence. Therefore clustering can be made more granular and instead of entire importance curves, segments of these curves can form the input to a clustering algorithm.

Visual encoding of an importance curve is informative on its own, but it can be further used as a basis for steering spatial visualization and animation. First of all, the clustering can be highlighted in a 3D view, showing what the cluster looks like structurally. Second, the playback of an animation sequence can be non-linearly modulated to prolong the duration of the most important events of the animation and shorten the duration of non-interesting segments of the animation sequence. Third, one can select a representative set of animation time steps. By calculating a cumulative importance function, each time steps importance $A(t_i)|t_i \in t_1, t_2, \ldots, t_n$ is summed up with importances of the previous time steps. Once the sum reaches a value of the given importance threshold, the sum is initiated with zero and the next time segment is being defined. For each time segment then, the highest importance representative

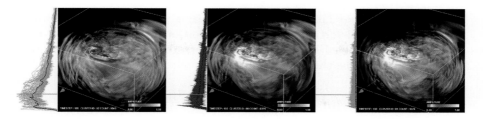

Figure 4.3 Three importance curve clusters have been identified in the Northridge earthquake dataset. The spatial view shows a particular time step depicted by the red horizontal line. The color encodes the region in the 3D view that is associated with the selected cluster in the time-importance plot, where the time is on the vertical axis and the importance on the horizontal axis (increasing time toward the top, increasing importance toward the left) [139].

is chosen to form the *storyboard* or set of representative time steps of the temporal sequence.

The importance-driven visualization is shown on an earthquake simulation dataset in Figure 4.3, where it essentially measures the earthquake's activity. It shows a peak importance at the early phase of the simulation. The importance curves are clustered into three regions based on the change in importance over time. The first cluster is located in the epicenter of the earthquake. The second cluster, which changes the importance values only mildly, surrounds the epicenter, and the constant-importance cluster is located on the periphery of the simulated dataset.

4.2   LEVEL-OF-DETAIL CHARACTERIZATION

Volumetric data representation is greedy in terms of memory or storage requirements. Full body medical scans, dynamic time-varying datasets, or datasets acquired during large-scale seismic surveys are still challenging to maintain in a form that allows for interactive data visualization. The most common way to deal with large data is the generation of a level-of-detail (LOD) data structure. In volume data this would typically be an octree that stores the original voxel values as leaves and the average value of eight of its siblings as intermediate nodes. Then a particular level-of-detail composition is essentially a pruned tree with each branch providing a specific level of detail. This pruned data structure will typically substantially reduce the amount of data. However, a natural question arises. How well does the reduced data reproduce the original data volume?

The following description is based on an *LOD-Map* toolset proposed by Wang et al. [137]. Each node of an LOD structure carries a certain level of distortion due to the data aggregation and value variation, while the leaves naturally have no distortion. The level of distortion d_{ij} of all the higher-level nodes can be quantified by assessing the differences between the values of a parent node i and the child node j:

$$d_{ij} = \sigma_{ij} \frac{\mu_i^2 + \mu_j^2 + c_1}{2\mu_i\mu_j + c_1} \frac{\sigma_i^2 + \sigma_j^2 + c_2}{2\sigma_i\sigma_j + c_2}, \tag{4.7}$$

where σ_{ij} is the *covariance* between the parent and the child node, which measures the strength of the correlation between them, and the second and third terms are the commonly used *luminance* and *contrast* distortions originating from the video analysis literature. These terms are based on the mean μ and standard deviation σ values of the parent and child nodes computed during the LOD buildup. The c_1 and c_2 are just small constants to avoid the numerical instability. The overall distortion D of a parent node i is defined as a sum of the distortions with respect to its children added with a maximal distortion of i's children with respect to its grandchildren:

$$D_i = \sum_{j=0}^{7} d_{ij} + max(D_j|_{j=0}^7). \tag{4.8}$$

From the volume visualization perspective, it is not only important how much distortion the intermediate node introduces in terms of data value, but also how the *visual appearance* of a reduced representation differs from the visual appearance of the original data. This is naturally dependent on the viewpoint and the transfer function settings. The *contribution* C of the node i on the visualization can be expressed as an approximation of the node's emission, calculated as a multiplication of mean value μ, average viewing ray length inside the segment l, and projected area of the intermediate node onto the viewpoint a. The approximated emission term is weighted by the visibility v of the node:

$$C_i = \mu \cdot l \cdot a \cdot v. \tag{4.9}$$

The properties *contribution* and *distortion* of a node in a multi-resolution LOD structure characterize a node particularly well and they can be used for quantifying the quality of a particular LOD composition. An LOD composition can be defined as good if the accuracy is well balanced over the volume and the contribution of high distortion areas is low. Here the information-theoretic measures come into play. Entropy, as the measure of average uncertainty of a random variable, offers a suitable measure. The random variable are the nodes, whose probability can be expressed as a normalized product of contribution C and distortion D

$$p_i = \frac{C_i \cdot D_i}{\sum_{j=1}^{M} C_j \cdot D_j}, \tag{4.10}$$

where the denominator is a sum over all M multiresolution nodes in the octree. The quality of an LOD composition is then computed by using Equation 1.1 for the entropy measure. The higher the entropy, the more balanced the overall composition is, which characterizes the LOD quality. Guided by such a quantitative measure of the LOD composition quality, the LOD can be further optimized. The strategy is to first join nodes with a very low probability. If the entropy does not decrease, then the *join* operation should be performed. Otherwise, the node is marked as processed and the next low-probability node is evaluated to see if it decreases the entropy. This is performed until all blocks are marked. Then, all the highest-probability nodes should be investigated. If *splitting* the highest-probability node would increase the entropy, then the splitting operation is performed. Otherwise, it is marked and the evaluation proceeds with the next high-probability node of the octree multiresolution structure. The third pass attempts a simultaneous join–split operation for one node with the lowest probability and one with the highest probability values. If the joining and splitting operations increase the entropy, then this double operation is performed. Otherwise, the pair is out and the next iteration is carried out.

While the entropy of a particular LOD composition is an expressive overall measure, it does not provide further details on how the particular quality level is achieved. For this, a compact visual representation called an *LOD map* has been designed. LOD map is based on a treemap, showing all nodes of a particular LOD composition as rectangles. The size of the rectangle is defined by the emission term of a node (the *contribution* without the visibility term), the visibility term determines the opacity of the block, and the color determines the distortion, where the red color encodes high distortion and the blue color encodes the low distortion. The LOD map can be analyzed during the interaction, when the viewpoint is changing or when the transfer function settings are modified. The interaction with the LOD map, such as splitting or joining of nodes, is also linked with the volume visualization. Clipping the volume with a plane will cause those nodes in the LOD map to become crossed. The visual encoding of an LOD map is in agreement with the holistic entropy measure and both are very suitable for comparing various LOD algorithms. This is shown in Figure 4.4, which shows the pelvis dataset represented by two different LOD compositions. While they have identical storage requirements, their quality largely varies. The left example is smoother, while the right one is richer in visual information. This is reflected both in the entropy values as well as in the LOD map encoding, where the left map shows significantly more distortion than the map on the right.

4.3   ISOSURFACES

The previous section presented one common strategy for handling large volumetric data by designing and assessing a level-of-detail scheme. In this section, another strategy is presented, i.e., isosurface extraction. The term *isosurface*

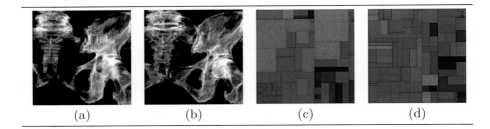

Figure 4.4 Two LOD compositions of a pelvis dataset consisting of 108 nodes each. The composition in (a) achieves entropy = 0.251 and (b), which is visually richer in detail obtained entropy = 0.414. The next two images show the corresponding LOD maps for lower entropy (c) and higher entropy (d) examples [137].

denotes a surface that is fitted to the volume so that it intersects only one value along the entire surface, namely its isovalue. An isosurface can be several orders of magnitude smaller than the original volume, and still represent a key structure within the volume, making isosurfaces a popular compressed representation. This strategy is very practical in acquired medical computer tomography scans for example, as a particular intensity value (or more precisely a *Hounsfield unit*) often relates to a specific type of a tissue. It is easily possible to find corresponding isovalues for skin or bones for example. In the volumetric data originating from simulation, particular isovalues may have a physical meaning, such as a particular density of a gaseous mixture, or pressure and temperature levels. Extracting informative isosurfaces, however, might be a time-consuming task, especially for identification of finer structures defined within a narrow band of the intensity spectrum. In time-critical scenarios, it would be very beneficial to obtain the set of most representative isosurfaces automatically. This section will describe how such process can benefit from utilization of information-theoretic measures.

The first described approach is known as *Isosurface Similarity Maps*, which has been introduced by Bruckner and Moeller [13]. The approach introduces a new matrix structure that shows which isovalues have similar corresponding isosurfaces and allows for intuitive visual exploration. Furthermore, the similarity matrix can be used as the basis for automatic isosurface extraction. The matrix stores, in each cell, the similarity value between isosurfaces corresponding to isovalues i and j. For each volumetric dataset, therefore, a time-intensive preprocessing computes how similar two isosurfaces are to each other. This is performed for all isovalue pairs to fill out the entire matrix. The matrix is symmetric with the highest values of similarity along its diagonal. As the measure of similarity, this approach uses normalized mutual information that spans the range $[0, 1]$. Mutual information expresses the degree of dependence of one random variable on another one, which makes it very suitable for expressing the similarity, analogous to earlier examples of use. The mutual information is computed from Equation 1.12 and the normalized mutual information uses the equation

$$\hat{I}(X;Y) = \frac{2I(X;Y)}{H(X) + H(Y)}. \tag{4.11}$$

The entropy H of each surface random variable is computed by the familiar formula from Equation 1.1.

The random variables X and Y are the distances from a point in space to the respective isosurface. The joint entropy is computed from the joint probability of these distance random variables. Each isosurface is first converted into a distance volume, which is a volume dataset, where each voxel stores information about the closest distance to the isosurface. Then a joint histogram is built from the two distances. For each voxel, a distance d to isosurface i and j forms the distance pair and the histogram will be incremented at a position $[d_i, d_j]$. The joint probability matrix is therefore a normalized 2D histogram, where the distance is on one axis to the isosurface i and the distance to the isosurface j is on the second axis. It is more precise to say that the histogram contains bins of distance intervals rather than specific distance values. The sum along one column or row, respectively, gives a marginal probability for one specific random variable and can be used for computation of entropy of these two random variables. Once the two marginal entropies and the joint entropy are calculated, the mutual information and the normalized mutual information are computed, which are then stored at a given place i, j of the isosurface similarity matrix SM, where V is the ordered set of isovalues. Moreover, summing up along row or column i in this matrix, an isosurface similarity distribution is obtained:

$$SD(i) = \frac{1}{|V|} \sum_{j=1}^{|V|} SM(i, j). \tag{4.12}$$

When plotting this distribution as a function plot over all isovalues, it becomes immediately apparent which isosurface intervals exhibit significant similarity. The peaks are the regions of high similarity while the valleys separate dissimilar isosurface intervals.

The isosurface similarity map can be utilized in several use case scenarios. In many applications of volume visualization, the intensity values of structures of interest vary slightly. In case of contrasted vascular scans, for example, it might be that some smaller vessels have, on average, a slightly lower intensity value than the larger vessels. To be able to visualize a complete vessel tree, it would be beneficial to take such intensity shift into consideration when visualizing or extracting the vascular structure. Computational analysis of the isosurface similarity map can identify similar isosurfaces in the same spatial region. For a particular position in space whose isovalue is different than the user-defined isovalue h_u the following importance function is evaluated:

$$\gamma(x) = \prod_{y \in C(x)} SM(h_u, f(y)), \tag{4.13}$$

where the $C(x)$ denotes the spatial neighborhood around point x. This $\gamma(x)$ is then used as a new opacity value instead of the original settings. In case of high similarity of the isosurfaces corresponding to the point x and its neighborhood $C(x)$, the visual representation of structure at location x will get a high opacity value.

Another application scenario deals with transitions between isosurfaces. During exploration of the volume data content, the user uses a slider to browse through the isosurfaces contained in the dataset. Such slider-triggered visualization can have a very non-linear response in visualization. For some intervals in the isovalue range, there could be almost no change in the visual appearance and then suddenly there is a drastic change just by small movement of the slider. With the similarity measure controlling the speed of the transition between isosurfaces, it is possible to achieve a more balanced and visually linear transition between the isosurfaces. A linear mapping ML from a slider position x on interval $[0, 1]$ to a given isovalue on interval $[h_{min}, h_{max}]$ is given as

$$ML(x) = h_{min} + x(h_{max} - h_{min}). \tag{4.14}$$

A similarity-based mapping MS should also be a monotonous function with $MS(0) = ML(0)$ and $MS(1) = ML(1)$, but the *speed* of the transition should be controlled by a *cumulative similarity* function SC:

$$SC(i) = \sum_{j=1}^{i} SM(j - 1, j). \tag{4.15}$$

The similarity-based mapping is defined as

$$MS(x) = ML(x) \frac{SC(\lceil x(|V| - 1) \rceil)}{SC(|V|)}, \tag{4.16}$$

where V represents the ordered set of isovalues.

The last example for application of isosurface similarity maps is the selection of the most representative isosurfaces. First, the isosurface with the highest average similarity value is selected. Then other values are selected that correspond to the peaks of the isosurface distribution function SD below and above the highest average similarity and stored in a priority queue. Once the set of maxima is collected, the first isovalue in the queue h_m is taken that initially corresponds to the highest average similarity isosurface and is selected as a representative isosurface. This value is removed from the queue and all the other isovalues whose surfaces exhibit high similarity to the surface corresponding to the h_m isovalue are moved backward in the queue. Then the first

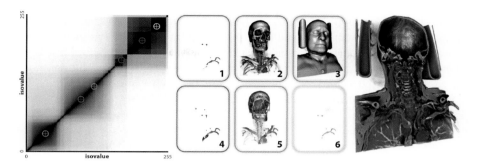

Figure 4.5 Representative isosurfaces of a volumetric dataset. The iso-surface similarity map on the left encodes by gray levels the amount of similarity among isosurfaces. The points along the diagonal are the isovalues forming the set of most representative isosurfaces, which are depicted in the middle. The clipping shows an integrated view of the six most representative isosurfaces [13].

isovalue in the priority queue is selected again for the next representative iso-surface and is removed from the queue. The isovalue reordering is performed again. This process continues with fewer and fewer representative isosurfaces added to the set of representative isosurfaces. The repositioning backward in the queue is guided by the following assignment:

$$p_i \leftarrow \frac{p_i}{1 + SM(h_m, h_i)}, \tag{4.17}$$

where the p_i is the position in the queue of the isovalue h_i and h_m is the isovalue of the first element in the queue.

The set of representative isosurfaces is shown in Figure 4.5. The under-lying data structure for the computation of the most representative surfaces for characterizing the content of a volumetric dataset is the isosurface simi-larity map, which is a symmetric matrix with highest values, in black, on the diagonal. The isovalues corresponding to the most representative isosurfaces are visually highlighted on the map. The most representative isosurfaces are related to the air–soft tissue interface in green, the bone tissue in blue, and the contrasted tubular structure in red.

There is another approach to finding a representative set of isosurfaces that communicates the overall content of a volume dataset and the level of detail, i.e., the number of isosurfaces can be quantitatively parameterized. This second approach is based on the method proposed by Wei et al. [142]. The es-sential idea of the approach is to find an initial set of isosurfaces—perhaps the first two based on minimal and maximal isovalues—to analyze whether a level-set-based blending between these two isosurfaces would closely correspond to the true isosurfaces between these two isovalues. The close correspondence is assessed by means of information theory. Blending between the two isosurfaces is realized by volume-based level sets. If an intermediate level-set surface closely matches a particular true isosurface between the two isovalues, then these marginal isosurfaces are good representatives for that isovalue interval. In case the match is weak, a new isosurface between the bounding two isosurfaces is added to increase the informativeness of the set of representative isosurfaces.

For any given polygonal surface embedded in a scalar field, the surface intersections with the scalar field can be computed. Also, the number of intersections of a particular value or a value interval can be summed up in a histogram. An isosurface that follows the distribution of one isovalue in the volume, intersects naturally just one single scalar value within the entire volume. This means that the histogram of intersected intensities would result in a high peak of one value only and zero occurrence of other intensity values. This process can be characterized by means of information theory, specifically by using the entropy measure. The scalar data value is our random variable. The probability of the random variable mapping to a particular value (or value interval) that is intersected by the surface can be expressed by the frequency of its intersections with the given surface. Based on the formula of the entropy measure, isosurfaces would sum to minimal entropy values, while any other surface that intersects more than one isovalue would result in a higher entropy value. Therefore, the closeness of the fit between two surfaces is evaluated by the entropy measure.

In the particular case of finding representative isosurfaces, the entropy value of an intermediate surface generated as a blend between two isosurfaces is evaluated as mentioned above. If the entropy value is largely different from the entropy value of its corresponding isosurface, then these are considered to be different. The intermediate surface is a blend generated by a volume-based level-set method. In simple terms, the two starting and ending isosurfaces are converted into a distance volume. Then every intermediate surface can be generated by an interpolation of these two distance volumes and a consecutive marching algorithm generates the intermediate surface. This surface is then assessed with respect to its closeness to an isosurface.

Within the interval between two isosurfaces, a set of intermediate surfaces is defined, which are interpolants, using the level-set method. For each level-set surface s_j, a histogram of frequency of intersecting particular isovalues x_i (or a value interval) is computed. In practice, as the triangulation is made by marching cubes algorithm with a very high vertex count, the vertices of the triangulated surface are at the same time the sample points of that surface. This forms the joint normalized histogram $h[]$, which is denoted as an *isosurface information map*. Turning the problem into an information theory scenario, the normalized version of the joint histogram is

$$p(x_i, s_j) = \frac{h[x_i, s_j]}{\sum_k \sum_l h[x_k, s_l]}. \tag{4.18}$$

The next step is to identify the least fitting level-set surface to the isosurfaces. This is based on the concept of a specific conditional entropy. Based on the joint probability matrix, marginal probabilities for both random variables are calculated by simply summing up rows or columns of the matrix. Specific conditional entropy $H(S|X = x_i)$ is then calculated based on Equation 1.4 and it expresses its distribution over the intermediate level-set surfaces:

$$H(S|X = x_i) = -\sum_{j=1}^{N_y} p(y_j|x_i) \log_2 p(y_j|x_i). \qquad (4.19)$$

A large entropy value means that this isovalue contributes to several of these surfaces, while a small value means that it is intersected by one or very few level-set surfaces. The isovalue with the highest specific conditional entropy is represented in the worst way from all the isovalues with the given interval.

It is also possible to evaluate the overall fit of all level-set surfaces to the isosurfaces within a data value interval. For this, the measure of conditional entropy between two random variables is especially suited as it measures the uncertainty between one random variable given the knowledge of another random variable. Conditional entropy of an isosurface random variable given the level-set surface random variable can be computed from the formula

$$H(X|S) = H(X) - I(X;S), \qquad (4.20)$$

where the entropy $H(X)$ is computed from the marginal distribution $p(x)$ and the mutual information $I(X;S)$ is computed from the joint probability $p(x,s)$ and the marginal distributions $p(x)$ and $p(s)$. The lower the conditional entropy is, the lower is the uncertainty about random variable X given the knowledge of random variable S.

In order to make the measures of conditional entropy and specific conditional entropy independent of the number of level-set surfaces, these measures are normalized by the theoretical maximal value they could reach, which is in case of $H'(X|S) : max = \log_2(N_x)$ and for the $H'(S|X = x_i) : max = \log_2(N_s)$.

These measures are used in the algorithm for finding the most representative set of isosurfaces as follows: First, the initial set of isosurfaces is determined. Either minimal and maximal isosurfaces form the initial set or the set is computed by another method. Then, for each interval, the level-set surfaces are computed and the conditional entropy $H(X|S)$ is calculated. If the value is above a certain threshold, this interval will be split and the splitting isovalue has to be identified. This is determined by the highest value of specific conditional entropy $H(S|X = x_i)$. The stop criteria for the recursive splitting are (1) the conditional entropy is below a certain threshold, or (2) the boundary isosurfaces are already very close to each other.

The technique is exemplified on several volumetric datasets including a single time step dataset from a simulation of thermal plumes. This dataset is

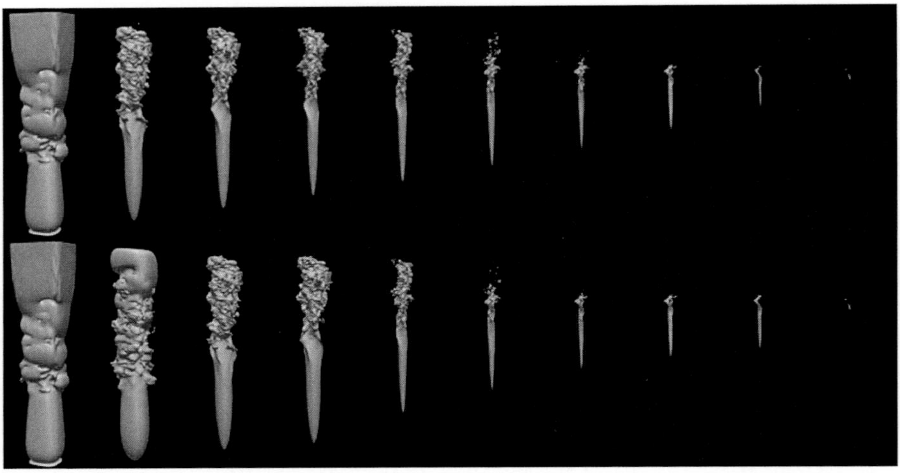

Figure 4.6 Representative isosurfaces of a simulated thermal plume dataset. The top row shows a selection of isosurfaces by regular sampling in the interval of permissible isovalues. The bottom row shows the representative isosurfaces picked by the isosurface information maps method. One can see that the second isosurface from the bottom left represents the continuous transition better and the uniform sampling does not provide enough shape cues to communicate structures at the low intensities end of the spectrum [142].

represented in Figure 4.6 showing two approaches of selecting representative isosurfaces. The top row shows uniform sampling for isovalue selection and the bottom row shows the selection using the isosurface information maps. The isosurface selection driven by information-theoretic measures provides more detail about the structural characteristics of the thermal plume over the naive uniform sampling approach.

4.4 SPLITTING

The previous section discussed one strategy for understanding structural characteristics of a complex volumetric dataset, by investigating the most prominent isosurfaces. Another approach for exploring the dense volumetric structure can be realized by various forms of clipping, splitting, or related exploded view techniques. The technique, known as exploded views, separates sub-components of the structure to see the internal parts more clearly. This technique is often used in technical drawings and can be adapted for volume data exploration in such a way that the data is separated into a set of volume regions that are internally very similar. The internal content of these subregions will be made visible by turning the volume with the splitting plane toward the viewer. Such splitting can be introduced at several spatial locations to reveal the structural details of internal structures, normally occluded by the outer layers.

The process of exploded views generation can be automated. As proposed by Ruiz et al. [100], the similarity-based exploded views method first automatically identifies the axis of explosion, which is the axis along which there is most of the structural change along the dataset. The second step of the explosion design is the identification of splitting planes that reveal more structural information as compared to other splitting plane positions. The selection of several splitting planes can be realized in a top-down manner by iterative splitting of the entire volume into smaller and smaller sub-volumes. The decomposition will terminate when there is very little dissimilarity within the volume slabs, or when the volume slab contains only a single slice. Another approach is a bottom-up strategy that starts with comparing individual slices as elementary building blocks of a volume, and those slices that are very similar in terms of the on-slice content can be merged together to form a volume slab. After several iterations, only a few clusters remain, which separate into largely different structures. These then form the sub-volumes that are placed apart from each other along the axis of explosion.

The necessary components for making the explosion design as an automated process can benefit from measures established in the information theory. The axis of explosion along the most structured direction can be estimated with the *entropy rate* measure. This measure is suitable for expressing the randomness or unpredictability of a sequence. In this case, the sequence is the order of intensity values along the ray of the potential explosion axis direction. The probability of a sequence of values can be expressed by a joint probability. The joint probability can be obtained from the joint histogram of occurrence of particular sequence values. Based on the joint probability, the entropy of the sequence is computed based on Equation 1.24. The L in this case is the length of the sequence consisting of consecutive samples along the ray. This joint probability can be used for computing the entropy rate measure as defined in Equation 1.25, which can be written as

$$h^x = \lim_{L \to \infty} (H(X^L) - H(X^{L-1})). \tag{4.21}$$

The computation of the entropy rate is realized by casting parallel viewing rays for each pixel into the volume. These rays are sampled at a constant rate and during the ray traversal, two joint histograms are built during the casting X^L and X^{L-1}. Once the entropy rate is computed, the direction of cast rays that has resulted in the highest entropy rate is selected as the view with the strongest structural changes along the ray. The view vector defines the axis of explosion.

After the axis of explosion is obtained, it is still necessary to compute the positions along which the volume will be split to reveal the internal structures. Two strategies, the top-down and the bottom-up ones, are explained in the following in detail. The top-down strategy is based on the maximum gain of information using the Jensen–Shannon divergence measure (Equation 1.30). The information is gained by splitting two regions that are not similar to each other and it is calculated as

$$\Delta I = \frac{v_1 + v_2}{v_T} JS(\frac{v_1}{v_1 + v_2}, \frac{v_2}{v_1 + v_2}; p_1, p_2), \qquad (4.22)$$

where v_1 and v_2 are the volumes of the slabs 1 and 2, v_T is the volume of the entire dataset, and the p_1 and p_2 distributions are the normalized intensity histograms of these two slabs.

The bottom-up approach is based on slice (or slab) similarity. For the similarity measure, the normalized mutual information measure is employed. First, all neighboring slices are examined for similarity by computing normalized mutual information. Then the slices with highest mutual information are merged into one slab and the mutual information between the new slab and the neighboring slices is recomputed. In the second iteration, the next two slices/slabs are merged into a slab, and the similarity is recomputed until either all slices/slabs are merged into the entire volume or a given dissimilarity among slabs is reached so that no merge can happen that would not exceed the dissimilarity threshold. The normalized mutual information $\hat{I}(X;Y)$ is calculated by the following formula:

$$\hat{I}(X;Y) = \frac{I(X;Y)}{H(X,Y)}, \qquad (4.23)$$

where $I(X;Y)$ is the mutual information from Equation 1.10 and $H(X,Y)$ is the joint entropy from Equation 1.3.

The entire presented pipeline is demonstrated in Figure 4.7, which shows the two different splitting/merging strategies. One can see that they do not deliver identical results and that the positioning of the splitting planes slightly differs. By turning the slab with the splitting plane toward the camera, the user can obtain a better understanding of the internal organization of a complex volumetric dataset.

4.5 TRANSFER FUNCTION DESIGN

The most common approach to display volumetric data is the direct volume rendering. The approach uses the transfer function that maps the scalar data values to a color and opacity value. This could be expressed as $f(\mathbf{x}) \rightarrow \{R, G, B, \alpha\}$, where $f(\mathbf{x})$ is the date value at spatial position $\mathbf{x}$. The optical properties $\{R, G, B, \alpha\}$ are then used during the ray casting and have a strong influence on how a particular voxel with its intensity value will be represented in the final image. Besides the voxel's own opacity value, the opacity values of preceding voxels along the viewing ray have a strong influence on the voxel's final appearance in the rendered image. In case of two-dimensional transfer functions, the gradient magnitude is used jointly with the data value

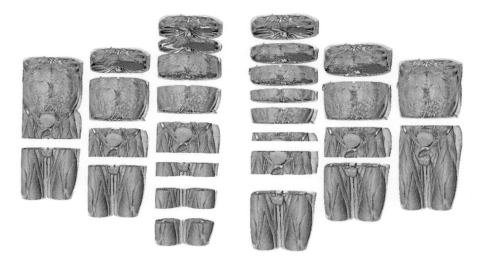

Figure 4.7 Exploded view of a full body scan with an automatically determined axis of explosion by the estimation of the most structured direction. Two techniques for volume splitting are used, the top-down approach (left) based on information gain via dissimilarity splitting and the bottom-up approach (right) based on the similarity evaluated with normalized mutual information [100].

as input for the transfer function mapping to the optical properties, which makes the transfer function even harder to design.

The definition of transfer functions can be done by manually defining the mapping from data to optical properties, however, this becomes a very tedious and time-consuming process. Moreover, even if the user has a clear idea of the voxels she would like to see, a transfer function manipulation does not intuitively lead to a desired visibility, as it globally modifies the visual appearance of all voxels. Therefore, the transfer function design, especially for non-expert users, is a lengthy trial-and-error process. A substantial research focus has been devoted to assisting the transfer function design. In this section, approaches on how to automatically estimate a good transfer function are described, given a definition of what the user wants to see and what is less important to her, so that the desired structures become visible on the resulting visualizations. The discussed approaches use the information theory in the automated search process.

The direct input to the transfer function are the data values (and associated local derivative properties). The final measurable outcome of visualization is the image rendered from a viewpoint that allows analysis of how a given voxel or a set of voxels with the same data value is visible. To characterize this transform, Ruiz et al. [100] propose to establish an information-theoretic channel between a set of viewpoints and binned data values. The channel captures the relationship between bin b and viewpoint v in a transition probability matrix

$$p(b|v) = \frac{vis(b|v)}{vis(v)}, \qquad (4.24)$$

where $vis(b|v)$ is the sum of visibilities from one viewpoint v of all voxels having the data value $f(\mathbf{x}) = b$. What the viewpoint sees from all data values is captured as $vis(v) = \sum_{b\in\mathbb{B}} vis(b|v)$, where $\mathbb{B}$ is the alphabet of binned data values. The visibility of a voxel is defined as the contribution of the classified voxel to the final image, according to its opacity and according to the accumulated opacity of voxels between the voxel at position $\mathbf{x}$ and the viewpoint. The rows in the transition probability matrix are normalized so that $p(B|v) = \sum_{b\in\mathbb{B}} p(b|v) = 1$.

The input distribution $p(V)$ contains the probability of each viewpoint

$$p(v) = \frac{vis(v)}{\sum_{i\in\mathbb{V}} vis(i)}, \qquad (4.25)$$

where the $\mathbb{V}$ is a set of viewpoints and $p(v)$ expresses a significance of a viewpoint. The output distribution $p(B)$ contains the probability of binned data values and it is calculated from the input distribution and the conditional probabilities:

$$p(b) = \sum_{v\in\mathbb{V}} p(v)p(b|v). \qquad (4.26)$$

The output distribution $p(B)$ describes an average projected visibility of binned data value b from all viewpoints. The information channel is essentially needed to capture $p(B)$. Next, the output distribution is related to a desired distribution of how data values should be visible on the resulting visualizations. The outcome of the alignment between the output and the desired distributions will control the transfer function modification.

The informational divergence (or the Kullback–Leibler distance) describes the divergence between two probability distributions: the true distribution p and a target distribution q. Informational divergence can be computed as described in Equation 1.12. For the automatic transfer function design, the informational divergence between the average projected visibility of binned data values $p(B)$ and some desired $q(B)$ should be minimized [98].

Various target distributions can be used to serve the purpose of automatic transfer function design. For example, a uniform distribution $q(B)$ can have a constant probability for each data value bin:

$$q(b) = |\mathbb{B}|^{-1}. \qquad (4.27)$$

Minimizing divergence to such a target distribution would mean that each data value bin is on average seen equally to the other data value bins. Another distribution can be histogram based. A set of voxels with the same binned data value b is defined as $S_b : \{\mathbf{x} \in S_b : f(\mathbf{x}) = b\}$. The histogram-based target distribution would mean that the more a particular value is present in the dataset, the higher should be its overall visibility in proportion to the other data values. Such a target distribution can be is defined as

$$q(b) = \frac{|S_b|}{\sum_{i \in \mathbb{B}} |S_i|}. \tag{4.28}$$

The third distribution can be based on the data value itself, which can be beneficial in situations when the important feature is marked with a high intensity value, such as the contrast agent in medical imaging. Its formulation is simply expressed as

$$q(b) = \frac{b}{\sum_{i \in \mathbb{B}} i}. \tag{4.29}$$

The fourth target distribution is extended to characterize the binned gradient magnitude:

$$q(b) = \frac{|\nabla f_b(\mathbf{x})|}{\sum_{i \in \mathbb{B}} |\nabla f_i(\mathbf{x})|}. \tag{4.30}$$

In such case, the binning is not only done based on the data value, but has a form of a joint histogram of data value and the data gradient magnitude. The transition probability matrix $p(B|V)$ uses binning where data values and gradient magnitude jointly form the bins. The calculation of $p(v)$ and $p(b)$ is as it was described earlier for the data-value-only scenario.

Furthermore, the target function can be further specified by the importance function that is assigned to segmented regions of the volume data. Voxels that belong to the segmentation mask will have the data value changed by adding one additional most significant bit to their fixed-point value range. Besides the fact that the number of bins is increased, the calculation of input and output distributions follows the above description of target distribution based on data value.

Once the target distribution $q(b)$ is specified, its divergence from the true distribution $p(b)$ can be estimated using the Kullback–Leibler distance. There are two possible formulations for informational divergence: a global one and a view-dependent one. The global one takes the entire distribution $p(B)$ into account, while the viewpoint-dependent one takes only visibility from one particular viewpoint into account: $p(B|v)$. This measure needs to be recomputed for every viewpoint change.

The data values and the data gradient magnitude are mapped by means of a transfer function to a certain opacity value (besides the color). The transfer function setting can be defined as a set of opacity values $A = \{\alpha_1, \alpha_2, \ldots \alpha_n\}$ and each such set is used during the image compositing in direct volume rendering. So the set A has a direct effect on the resulting value of the informational divergence measure. The search for an optimal transfer function is turned into an optimization problem with an objective function

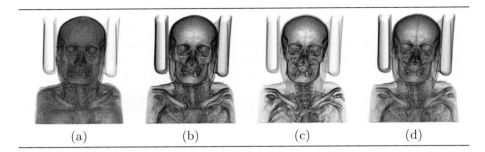

Figure 4.8 Automatic transfer function design based on the global strategy, (a) histogram-based target distribution, (b) histogram and gradient based, (c) importance assigned to the bone tissue and gradient-based target distribution, and (d) single viewpoint strategy with histogram- and gradient-based target distribution [98].

$F(A) = D_{KL}(p||q)$, which should be minimized using the gradient descent approach. Therefore, a new iteration of the transfer function optimization process is controlled by the formula

$$A_{t+1} = A_t - \nabla D_{KL}(p||q). \tag{4.31}$$

This means that a derivative of the informational divergence function needs to be computed. It can be shown that the derivative of the informational divergence can be approximated by

$$\frac{\partial D_{KL}(p||q)}{\partial \alpha_b} \approx \frac{p(b)}{\alpha_i(1 - p(b))} \left(\log \left(\frac{p(b)}{q(b)} \right) - D_{KL}(p||q) \right), \tag{4.32}$$

where b is a particular binned data value and α_b is the average opacity corresponding to this data value range.

After several iterations of the optimization process, the opacity values are adjusted to match the particular target distribution. This automated optimization process does not include the color design; the colors need to be assigned to the data values in a separate manual step. For medical scans, such as computer tomography scans, the tissue can be assigned a color that is a standard color for that tissue in medical textbooks. The results of the optimization process are shown in Figure 4.8 on a computer tomography scan with varying tissue intensities and strong gradients between the air–soft tissue interface and the soft tissue–hard tissue interface. One can see how the results differ based on the target distribution settings and the global and viewpoint-based strategy.

The automatic transfer function design above was driven by the average visibility of the binned data value and the relationship between the average data visibility distribution and some target, desired distribution. Another approach for establishing the relationship between input and output values of a visualization pipeline can be based on final pixel colors instead of voxel visibility for the output distribution. An *observation* channel was proposed by Bramon et al. [11] between an input data value and the output pixel, which can also be described by the conditional probability matrix. In this relationship, the visualization quality increases when the relationship between data values and final pixel colors is less uncertain. If too many voxels contribute to the final pixel color, it is unclear what the final projection represents. Therefore, the clearer this relationship is the better. This visualization quality measure can be applied on the automatic transfer function design, but also on viewpoint selection or light placement design.

The probability distributions of the input and the output variables can be obtained from the normalized histogram of binned data values $p(X)$ and binned pixel colors $p(Y)$. To compute the necessary probability distributions for the transitional probability matrix that is representing the observation channel, it is necessary to perform the ray casting twice. First, the ray casting is performed to compute the final pixel color value y. This basically describes the column in the transitional probability matrix to which the particular ray contributes. For each sample with data value x along the ray, the particular row representing the input data value x is incremented in a joint histogram for a pair composed of a data value and a color. This is done for all samples along the ray and for all rays (and possibly also for all viewpoints). Based on the joint histogram and the input probabilities, the transitional probability matrix $p(Y|X)$ of the observation channel is computed. From all the channel probability distributions, the mutual information (Equation 1.11) value can be computed that expresses the degree of dependence of the output pixel on the input data value. The higher this dependence is, the clearer the visualization communicates the underlying data values.

The transfer function optimization can therefore be based on the evaluation of the highest mutual information value among a set of candidate transfer functions. The opacity transfer function can also be expressed analytically, by a set of Gaussian functions G_i that are defined by their mean μ_i, variance σ_i, and amplitude φ_i. These Gaussian functions are summed up for each position of the intensity range to define the opacity value. For the transfer function defined by a particular set of Gaussian functions, the ray casting is performed. Based on the approach delineated above, the visualization quality is estimated by computing the mutual information value. This is initially done for a set of transfer functions and the one with the best visualization quality score is taken to the next iteration of the automatic transfer function design. The winning function is modified by perturbation of its parameters μ, σ, and φ to create a set of new transfer function candidates. Again, the best transfer function passes into the next round. This evolutionary optimization loop is terminated when, for several generations, there is no improvement in the visualization quality. Then the transfer function with the best visualization

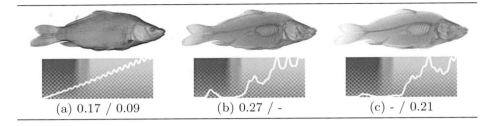

(a) 0.17 / 0.09	(b) 0.27 / -	(c) - / 0.21

Figure 4.9 Transfer function based on a set of Gaussian functions evolving based on the visualization quality measure (mutual information) between input data values and output pixels. The initial function approximates a linear ramp and its visualization quality value based on one view and six views (a), the best transfer function obtained from a one-view evaluation, and (b) the best transfer function obtained from a six-view evaluation (c) [11].

quality value is selected to represent the data. An example of the optimization process is shown on the computer tomography dataset of a carp in Figure 4.9. The initial transfer function shows one isosurface of the animal's surface. The transfer functions preferred by the visualization quality measure also shows the internal arrangement of the anatomical structures.

While the transfer function design is a strong use case for the observation channel, its potential use can be found for other optimizations of the visualization parameters. For example, when computing the mutual information visualization quality measure for each viewpoint around the volumetric dataset, the one that gives the highest quality value, can be considered as the best one. The best viewpoints are those where the outcome color is the most certain one.

The application usage for the best light placement is also done on the basis of the highest achievable mutual information value. The approach supports several light sources placed in the scene and the optimization scheme is as follows: The scene is initially illuminated with a light source placed in the camera position. The second light is identified by evaluating the increase of the visualization quality value. This is repeated for as many lights as desired or until there is no significant increase in the visualization quality.

The concept of voxel mutual information (VOMI) presented by Ruiz et al. [99] and discussed in the previous chapter (Section 3.3.2) can be interpreted as a measure of importance to modulate the opacity of a transfer function, in a kind of focus+context strategy. The focus of interest is considered as the most informative part of the volume. Then, the opacity of the most informative voxels is increased, or preserved, while the opacity of the least informative is reduced. This opacity modulation is driven by the equation

$$A'(z) = \begin{cases} A(z)k_\mathrm{l}\,\overline{I(z;V)}, & \text{if } \overline{I(z;V)} < t_\mathrm{l}, \\ A(z)k_\mathrm{h}\,\overline{I(z;V)}, & \text{if } \overline{I(z;V)} > t_\mathrm{h}, \\ A(z), & \text{otherwise,} \end{cases} \qquad (4.33)$$

where $A(z)$ is the opacity of the voxel z before modulation, t_l and t_h are the low and high thresholds respectively, k_l and k_h are factors that regulate the effect of the modulation, $\overline{I(z;V)}$ is the normalized VOMI, and $A'(z)$ is the opacity of the voxel after modulation.

In the following figures, VOMI is computed using 162 viewpoints. Figure 4.10 has been obtained by modifying the thresholds and factors in order to emphasize a selected part of the model while preserving the context. Figures 4.10(a) and 4.10(d) correspond to the original CT-body with $t_\mathrm{l} = 0$ and $t_\mathrm{h} = 1$ viewed from different viewpoints and with different transfer functions. In Figures 4.10(b) and 4.10(c) our target is the skeleton. The skeleton has a high VOMI because it is highly occluded and we have to decrease the opacity of the less-occluded parts, such as muscles, which have low VOMI. In Figure 4.10(b), the parameters are $t_\mathrm{l} = 0.5$ and $t_\mathrm{h} = k_\mathrm{l} = k_\mathrm{h} = 1$. In Figure 4.10(c), a more extreme effect can be obtained by setting k_l to 0.5, making less-occluded parts even more transparent. In Figure 4.10(e) the focus is on the ribs, making the muscles around them more transparent. This is obtained with $t_\mathrm{l} = 0.3$, $t_\mathrm{h} = 1$, $k_\mathrm{l} = 0.1$, and $k_\mathrm{h} = 1$.

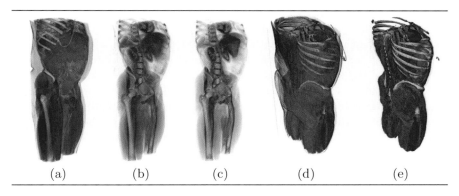

(a)	(b)	(c)	(d)	(e)

Figure 4.10 CT-body model visualized with different transfer functions: (a) and (d) in their original state, (b) and (c) modulated from (a) by a VOMI map to emphasize the skeleton, (e) modulated from (d) by a VOMI map to emphasize the ribs [99].

VOMI depends on the transfer function used to visualize the model. In the previous example, the transfer function of the volume is modulated with the VOMI computed with the same transfer function. However, it is also possible

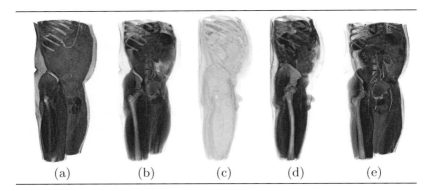

(a)   (b)   (c)   (d)   (e)

Figure 4.11 CT-body model visualized with the transfer function: (a) of the original model, (b) modulated by a VOMI map computed with the transfer function used in Figure 4.10(a), (d) and (e) modulated by a VOMI map computed from the viewpoint in (c) [99].

to compute the VOMI with one transfer function and use it to modulate another transfer function over the same model. To illustrate this possibility, the VOMI computed with the transfer function from Figure 4.10(a) is used to modulate the transfer function in Figure 4.11(a). Figure 4.11(b) is obtained by setting the parameters to the same values as in Figure 4.10(b). Since muscle in Figure 4.10(a) is more transparent than in Figure 4.11(a), modulating the opacity of the latter with the VOMI map of the former makes the muscle more transparent than it would be with its own VOMI map.

Instead of computing the VOMI from a set of viewpoints uniformly distributed over the surface of a sphere, it is also possible to compute it from just one of the viewpoints and a neighborhood. For instance, in order to emphasize the right hip and the femur, the VOMI map is computed considering the viewpoint at the right side of the model. Figure 4.11(c) shows the VOMI map, and Figures 4.11(d) and 4.11(e) show the modulation of Figure 4.11(a) obtained using this map.

4.6 MULTIMODAL VOLUME VISUALIZATION

Each imaging modality that produces a volumetric dataset shows one physical property of the studied phenomenon. Analyzing the same structure using several modalities simultaneously is an opportunity to gain a more complete understanding as compared to the analysis of a single modality at a time. One example from medical or industrial imaging are the so-called dual energy scanning procedures. Here, two volumetric scans are produced with two different energy radiation levels. While the lower energy scan features much higher detail, it suffers from artifacts as the low energy radiation does not penetrate through certain structures and is maximally attenuated. This creates areas that are imaged with many fewer projections than other areas and leads to artifacts such as beam hardening and the like. On the other hand, the higher-energy X-ray radiation penetrates through all imaged structures, but the resolution is lower. Combining these two energies can offer high detail with no attenuation artifacts. However in essence, both scans originate from the same modality based on X-ray radiation, so they image the same characteristics of the structure, namely the tissue density or the attenuation strength of the X-ray beam by tissue. In this case, we can say that one modality *supplements* the other one to eliminate certain modality-specific artifacts.

Another example from medical imaging, especially in the neurosurgical context, are multimodal imaging protocols that combine fundamentally different imaged characteristics. It is quite common to combine CT with MRI scans as the CT modality preserves original distances in structure and has a strong contrast between hard and soft tissues, while the MRI features strong contrast between soft tissues, such as the brain parenchyma. These two structural modalities can be combined with f-MRI or a PET scan imaging physiology on a macroscopic level and *complementing* the structural information with information about function. These scans are often performed separately and it is necessary to bring them to the common reference frame. For this purpose, the medical image registration methodology is utilized. The registration technique finds a transformation of one modality so that it perfectly matches the other modality. In this way, it is possible to understand the relationship of values from multiple modalities corresponding to the same voxel.

While having a more comprehensive protocol including several scanning modalities is definitely advantageous in terms of data quality, it comes with an additional challenge of how to represent this data. When analyzing the data by viewing slices, one modality can be represented by chromatic contrast, while the second one is encoded by a contrast in luminance. These two representations can be superimposed over each other and both modalities can be analyzed in the same spatial reference simultaneously. With 3D volume visualization, the situation is more complicated and Cai and Sakas [16] propose three levels of modality intermixing. The image-level intermixing renders each volume with separate optical properties with the same camera parameters. The final rendered images are fused together. This naturally leads to spatial inconsistencies in many cases. Accummulation-level intermixing alleviates this problem and uses two different optical properties for each modality, which are intermixed for each sample along the viewing ray. One stage earlier is the illumination-level intermixing, where optical properties are assigned to each combination of data values from the two modalities. In the following, we present various intermixing strategies that take advantage of information theory in the fusion process.

While not included in the Cai and Sakas categorization, intermixing can be realized on the data level as well. This means that prior to any classification, the data fusion is performed, and once a single scalar dataset is created, it can be visually analyzed as a *single* modality. The following method presented by Haidacher et al. [58] takes the data-level fusion strategy and uses information theory in the fusion process. First, a joint dual histogram of value occurrences is computed for a particular multimodal volume-data pair. Normalizing the histogram by the total number of voxels represents a joint probability $p(x, y)$ of a pair of values. Marginal probabilities of a single modality $p(x)$ or $p(y)$ can be computed by summing up rows or columns of such a probability matrix. In information theory, the *information content* of a particular value can be computed from the probability as $i(x) = -\log_2 p(x)$. The formula for information content is used as a weight term in a linear interpolation fusion process of the two input modalities. The weight γ is computed as a ratio of the information content between two modalities for a particular position in the volume

$$\gamma(x, y) = \frac{i(x)}{i(x) + i(y)}. \tag{4.34}$$

The fusion is then carried out by a simple linear interpolation between the two values $z = x(1 - \gamma(x, y)) + y\gamma(x, y)$. The gradient value of the fused function ∇z is computed by linear interpolation of gradients in the original scalar fields weighted by the $\gamma(x, y)$. Having the new fused quantities, the user can specify the optical properties using a standard two-dimensional transfer function widget with the fused value along the horizontal axis and the fused gradient magnitude along the vertical axis.

In addition to the fused value and the gradient vector, one more quantity δ is computed per spatial position. This quantity expresses how much the characteristic at the particular spatial position in one modality is expressed in the second modality and how informative it is. It can be computed from an information-theoretic quantity called the *pointwise mutual information* (PMI) defined as

$$PMI(x, y) = \log_2\left(\frac{p(x, y)}{p(x)p(y)}\right). \tag{4.35}$$

The PMI quantity is equal to zero when the combination of values occurs as frequently as would be expected by chance. This occurs when the two random variables are independent of each other. If this value is higher than zero, they occur more frequently than one would expect by chance and when the PMI is less than zero, the pair of values occurs less frequently than would be expected by chance. In terms of information content, this is the most informative occurrence. The new quantity δ measures the opposite information content and its value is inverted normalized pointwise mutual information, $\delta = 1 - \hat{PMI}(x, y)$. When this quantity is close to 1, it means that one modality conveys high-value complementary information as compared to another modality. One way to incorporate the opposite information content δ into the transfer function design is to extend the transfer function by one dimension into 3D. However, to simplify the user interaction, it is better to use it as a separately specified windowing function, which is multiplied by the opacity of the given voxel.

The results of the fused volume based on information content are shown in

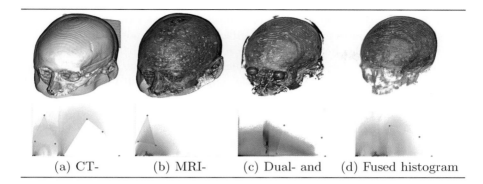

(a) CT- (b) MRI- (c) Dual- and (d) Fused histogram

Figure 4.12 The comparison of single modality volume visualization (a) and (b) and multimodal volume visualizations (c) and (d). The brain parenchyma is more clearly separated with data-level intermixing than with the illumination-level intermixing by means of the dual histogram. Figures (a), (b), and (d) use value vs. gradient magnitude histograms and (c) uses the value 1 vs. value 2 in the dual histogram [58].

Figure 4.12, which shows the two input modalities CT and MRI, visualization using a dual histogram of two function values, and the fused volume classified with a two-dimensional transfer function. The fused representation can separate the brain matter tissue from other structures much better as compared to the dual-histogram-based classification.

An example of a hybrid technique between data-level and illumination-level fusion has been presented by Bramon et al. [10]. The data fusion in their technique assigns a value to the fused voxel from the first modality or the second modality, by the winner-take-all principle. This has an advantage that no novel values are created that might be difficult to interpret, rather, the fused volume is constructed by parts from both modalities. For example, for head scans, the brain would be taken from the MRI while the rest of the anatomy would originate from the CT. An open question is which value should be chosen at a particular spatial location. For this purpose, the concept of the information-theoretic communication channel between two random variable distributions is utilized, where the mutual information describes the knowledge of one random variable about the second random variable. The communication channel can be built from the histograms of occurrences of particular values in the datasets of multiple modalities. A marginal probability for a particular value x, $p(x) = n(x)/N$ is equal to the number of the voxels with the value x normalized by the total of all voxels in the dataset. The values for the conditional probability matrix are computed as $p(y|x) = n(x, y)/n(x)$.

Once the communication channel is established, it can be analyzed and characterized by mutual information and its various decompositions, which is also called the *specific information* of x, described in Section 1.3. Among such specific information of x, the $I_1(x; Y)$ decomposition, known as the *surprise* (Equation 1.18), can be one candidate to characterize the understanding of Y from the value x. I_1 gets large values when the value for $p(y|x)$ is significantly higher than $p(y)$. Another decomposition of mutual information is I_2, also called *predictability* (Equation 1.19), which quantifies the change in uncertainty of Y when value x is observed. One more decomposition worth consideration is I_3, also called *entanglement*, which indicates that the most informative values of x are the ones that are related to the most informative outputs of y.

These decompositions of mutual information can be used for deciding which dataset to pick from the two modalities to characterize a particular spatial position. Essentially, the value z of the fused dataset is made according to the conditional formula in Equation 4.36:

$$z = \begin{cases} x, & \text{if } I_1(x; Y) > I_1(y; X) \\ y, & \text{otherwise} \end{cases}, \tag{4.36}$$

where the *surprise* decomposition was used as a fusion criterion. The criterion can be replaced by a another decomposition, such as *predictability* or other meaningful decompositions. Although at the first glance counterintuitive, a very effective criterion for modality selection is an asymmetric condition described in Equation 4.37:

$$z = \begin{cases} x, & \text{if } I_2(x; Y) > I_3(x; Y) \\ y, & \text{otherwise} \end{cases}. \tag{4.37}$$

The condition essentially indicates that in the first case, x is more informative than y. The complementary condition indicates that the values of y that are related to x are more informative than x.

Once the per-voxel selection using the specific information criteria is made, a new fused dataset can be displayed using ray casting. The sampling and gradient calculation are performed in the original single-modality volumes and for each sample along the ray, the decision is made whether the sample is taken from the first or from the second modality. The values of the I_2 decomposition can be seen in Figure 4.13, where several modalities are related to each other and for each modality the decomposition is computed for each value. We can observe, encoded by the thermal scale, that the *predictability* of the brain tissue parenchyma is fairly low for the CT, while it is rather high for the MRI-T1 protocol. This means that during the fusion process, the MRI values would be preferred over the CT values intuitively because the MRI offers much better contrast in brain tissue than the CT. Furthermore, we can see that in the fusion of the MRI T1 and T2 protocols, the T2 has a very strong value of the decomposition in the brain tumor region, estimating that it as a very informative part of the dataset. In case of the PET-CT fusion, we can see that

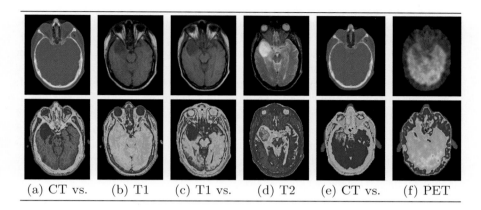

(a) CT vs.   (b) T1   (c) T1 vs.   (d) T2   (e) CT vs.   (f) PET

Figure 4.13   Information maps encoded with a thermal scale show specific information I_2 with respect to another modality. Blue regions carry low information content while the green and red regions are rich in information [10].

the bone tissue receives higher values from the CT than from the PET, and the brain activity imaged by the PET modality is selected for representing the brain parenchyma.

The hybrid fusion approach uses two transfer functions that are combined during the ray traversal. There is, however, no guarantee that the combination of the transfer functions will display the structures in a desired way. Bramon et al. [12] therefore extend their modality selection approach using specific information so that the optical properties per voxel are fused and the transfer function is optimized in the spirit of the optimization approach [98] described in Section 4.5 of this chapter. The process is initiated with two modalities and two good transfer functions, one for each modality. First, the relationship between these two modalities is characterized by an information-theoretic communication channel and studied by the specific information of x obtained from the decomposition of mutual information, similar to the approach discussed above. The condition used for fusion is the asymmetric one, described by the conditional formula in Equation 4.37. In the case of the transfer function adjustment, however, the color is defined by interpolation of the multimodal pair of values x and y. The color interpolation is described as

$$c(x, y) = \frac{I_2(x, Y)c_X(x) + I_3(x, Y)c_Y(y)}{I_2(x, Y) + I_3(x, Y)}, \qquad (4.38)$$

where I_2 and I_3 are the decompositions of the mutual information and c_X and c_Y are color transfer functions corresponding to each modality and $c(x, y)$ is the fused color for the value pair.

Once the color components are fused, the opacity channel is fused in the same way as the color. The same approach is used to fuse the gradient vector used in the illumination calculation. The fact that color, opacity, and gradient are fused for each voxel, makes the approach an illumination-level fusion approach. While the color and gradient are then determined and no longer modified, the fused opacity serves as an initial value that is further optimized. The optimization of the opacity value is performed by evaluating the alignment of a pre-defined target distribution q and the true distribution p. The true distribution is the average visibility of data bins from a given set of viewpoints calculated from the information channel established between the viewpoints and the binned data values. The target distribution can follow various characteristics, such as the occurrence of values in the dataset, meaning that the isovalue pair, which is assigned to most of the voxels, gets the highest visibility, while the lowest amounts receive the least visibility. More target distributions are described in Section 4.5. The data bins in the original method consisted of a quantized intensity range or the binning was a two-dimensional binning of pairs of intensity and the gradient magnitude. With two modalities, the binning is either two-dimensional when the two intensity pairs are binned into a set of bins, or three dimensional, when the fused gradient magnitude is considered for binning.

After binning, the optimization process proceeds as described in Section 4.5 by evaluating the informational divergence between the true and target distributions. The steepest gradient descent optimizer is utilized to find the next opacity distribution that better matches the true and target distributions. Once the match between these two distributions is below a certain threshold value, the final opacity value for a particular tuple of values and gradient magnitude is determined. Figure 4.14 shows single modalities represented using the modality-specific transfer functions. Further images show the optimized fused representations using various target distributions. One can see that in the multimodal visualization case, the pathology becomes very apparent as compared to the visualizations of single modalities of CT and MRI-T1.

The next illumination-level intermixing technique, *volume analysis using multimodal surface similarity* proposed by Haidacher et al. [57], extends the isosurface similarity maps approach described in Section 4.3 of this chapter. Having a two-modality volume set co-registered, such as a dual-energy CT scan or an MRI scan co-registered with a CT scan, a distance volume for each scalar value x and y representing an isosurface in each of the two modalities is computed. Then, for every pair of distance volumes originating from the modalities, a joint histogram of distance pairs is computed. The distance pairs, are the coordinates in the joint histogram. This histogram, when normalized, represents the probability that a given spatial point p has a distance $d(p, s_x)$ and $d(p, s_y)$ to the isosurfaces s_x and s_y. Based on this joint probability distribution, the mutual information and marginal entropies are calculated analogously to the isosurface similarity maps [13] to finally compute the normalized mutual information from Equation 4.11. The normalized mutual information is calculated for all scalar value pairs x, y to build a multimodal

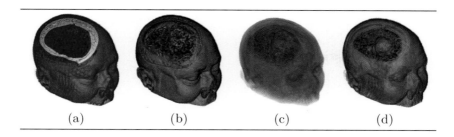

| (a) | (b) | (c) | (d) |

Figure 4.14 Direct volume rendering of (a) CT and (b) MRI modalities and (c) multimodal visualization using value pair occurrence as the target distribution and (d) value pair + gradient occurrence for describing the target distribution. One can see that the last image shows most of the structural detail, especially in the pathological area [12].

similarity matrix $MSM(x, y)$ that captures the similarity between the two input volumes. In contrast to the isosurface similarity matrix SM (Section 4.3), this matrix is not a square matrix, as each modality has in general a different range of data values. It is also no longer a symmetric matrix as the values $MSM(x, y)$ and $MSM(y, x)$ are not identical.

The multimodal similarity matrix can be utilized for data exploration. A typical approach to explore multimodal data is to use a dual histogram where each axis represents one modality or one of the techniques described above. The MSM is an alternative exploration tool where both similar and dissimilar structures become quickly discernible. The similarity between modalities can be, for example, directly controllled voxel opacities. In the multimodal context it can be valuable to analyze which isovalue in one modality best matches the isovalue in another modality. While with standard exploration tools this information is difficult to obtain, using the MSM, the highest similar isovalue in the *column* modality to a particular isovalue k in the *row* modality can be identified by simply finding the maximal value in the k-th row of the multimodal similarity matrix.

Especially useful is the multimodal similarity matrix for the classification purposes. The idea is that the user specifies optical properties for a small set of data-value pairs x, y and the classifier propagates the optical properties from pre-set data value pairs to all other unclassified pairs based on their similarity. For a given unclassified pair k, l, Equation 4.39 defines the classification process by which the most similar classified pair x, y is assigned to the unclassified pair k, l:

$$m(k, l) = \operatorname*{argmax}_{x,y}\big(MSM(k, y)MSM(x, l)\big). \tag{4.39}$$

Then each unclassified pair is assigned the optical properties of the most similar classified pair.

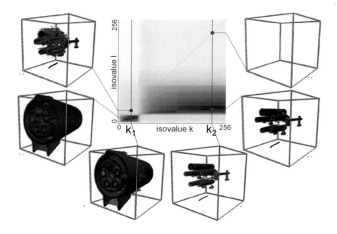

Figure 4.15   The multimodal similarity matrix shows strongly similar values with respect to their corresponding isosurfaces in dark shades of gray while dissimilar isosurfaces are in lighter gray levels. When in one modality isovalues corresponding to specific isosurfaces are known, the user can assume a linear relationship in the second modality as well. The red colored isosurfaces are the reference ones and the top row shows isosurfaces corresponding to manual selection assuming the linear relationship, which is however not the case. The middle row shows the isosurfaces determined by the MSM analysis as the closest to the reference isosurfaces [57].

The use of the multimodal similarity matrix is exemplified in Figure 4.15. Here a dual energy industrial CT produces two volumes with different radiation energies. An open question is how these two relate to each other with respect to their isosurfaces. When an isosurface determined by a particular isovalue k is known in one modality, the corresponding isovalue l in the second modality can be determined as

$$l = \operatorname*{argmax}_{y} MSM(k, y). \qquad (4.40)$$

The red isosurfaces represent the known isosurfaces from the first modality and the blue representation in the middle row shows the automatically identified corresponding isosurfaces in the second modality. The top row shows naive guesses for isosurfaces when attempting to find the isovalues manually.

4.7   SUMMARY

This chapter has provided an extensive overview of how information-theoretic measures can be utilized in addressing challenges associated with volume visualization. First, the focus was on time-varying volumetric datasets, where IT measures assisted the in selection of the best viewpoint, design of a comprehensive camera path, or which structures are interesting to look at based on their temporal evolution. IT measures found their usefulness in the level of detail design with a favorable trade-off between size and quality, in the selection of most interesting isosurfaces or in the design of a volume splitting so that the most interesting internal structures can be visually inspected. The last discussed topics were automatic transfer function specification and multimodal fusion driven by the quantities obtained from the information-theoretic measures.

FURTHER READING

Bordoloi, U.D. and Shen, H.-W. (2005). View selection for volume rendering. In *IEEE Visualization 2005*, pages 487–494, IEEE, Minneapolis.

Bramon, R., Boada, I., Bardera, A., Rodriguez, J., Feixas, M., Puig, J., and Sbert, M. (2012). Multimodal data fusion based on mutual information. *IEEE Transactions on Visualization and Computer Graphics*, 18(9):1574–1587.

Bramon, R., Ruiz, M., Bardera, A., Boada, I., Feixas, M., and Sbert, M. (2013). Information theory-based automatic multimodal transfer function design. *IEEE Journal of Biomedical and Health Informatics*, 17(4):870–880.

Bramon, R., Ruiz, M., Bardera, A., Boada, I., Feixas, M., and Sbert, M. (2013). An information-theoretic observation channel for volume visualization. *Computer Graphics Forum*, 32(3):411–420.

Bruckner, S. and Möller, T. (2010). Isosurface similarity maps. *Computer Graphics Forum*, 29(3):773–782.

Haidacher, M., Bruckner, S., and Gröller, M.E. (2011). Volume analysis using multimodal surface similarity. *IEEE Transactions on Visualization and Computer Graphics*, 17(12):1969–1978.

Haidacher, M., Bruckner, S., Kanitsar, A., and Gröller, M.E. (2008). Information-based transfer functions for multimodal visualization. *Proceedings of EG VCBM*, 101–108.

Ji, G. and Shen, H.-W. (2006). Dynamic view selection for time-varying volumes. *IEEE Transactions on Visualization and Computer Graphics*, 12(5):1109–1116.

Ruiz, M., Bardera, A., Boada, I., Viola, I., Feixas, M., and Sbert, M. (2011). Automatic transfer functions based on informational divergence. *IEEE Transactions on Visualization and Computer Graphics*, 17(12):1932–1941.

Ruiz, M., Boada, I., Feixas, M., and Sbert, M. (2010). Viewpoint information channel for illustrative volume rendering. *Computers and Graphics*, 34(4):351–360.

Ruiz, M., Viola, I., Boada, I., Bruckner, S., Feixas, M., and Sbert, M. (2008). Similarity-based exploded views. *Proceedings of Smart Graphics'08*, 154–165.

Wang, C. and Shen, H.-W. (2006). LOD map: A visual interface for navigating multiresolution volume visualization. *IEEE Transactions on Visualization and Computer Graphics*, 12(5):1029–1036.

Wang, C., Yu, H., and Ma, K.-L. (2008). Importance-driven time-varying data visualization. *IEEE Transactions on Visualization and Computer Graphics*, 14(6):1547–1554.

Wei, T.-H., Lee, T.-Y., and Shen, H.-W. (2013). Evaluating isosurfaces with level-set-based information maps. *Computer Graphics Forum*, 32(3):1–10.

Flow Visualization

CONTENTS

Fluid flow simulations play an important role in many scientific disciplines, for example, in the modeling of thermal hydraulics in a nuclear reactor. To visualize vector data generated from these simulations, one popular method is to display flow lines, such as streamlines or pathlines, computed from numerical integration. The primary challenge of displaying flow lines, however, is to place the particle seeds in appropriate positions such that the resulting flow lines can capture important flow features without cluttering the display. Setting parameters for flow visualization algorithms so that no important features are

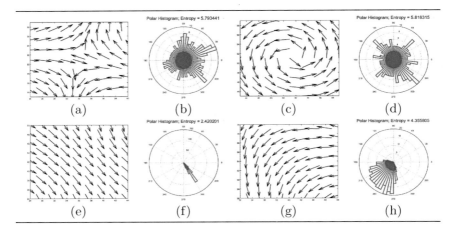

Figure 5.1 Distributions of vector directions. (a) and (c) are vector fields that have a saddle and a spiral source. (e) and (g) are two vector fields without critical points. The polar histograms with 60 bins created are shown in figures (b),(d),(f),(h). The entropy values of the vectors in (a), (c), (e), and (g) are 5.79, 5.82, 2.42, and 4.36, respectively. The range of the entropy with 60 bins is $[0, \log(60) = 5.9]$ [146].

missed is non-trivial. Furthermore, it is difficult to evaluate the quality of the visualization output. For large-scale datasets, scientists will not be able to set visualization parameters through trial-and-error because the cost of recomputing visualizations will be too expensive. To generate a visualization that can capture a maximum amount of information from the data with a minimal amount of visual clutter and user intervention, it is crucial to have automatic visual analysis algorithms combined with quantitative quality measures. In this chapter, we introduce several information-theoretic measures and their related algorithms to tackle this problem.

5.1 COMPLEXITY MEASURE OF VECTOR FIELDS

We can measure the complexity of vectors in a local region by analyzing the distribution of data. The underlying idea is that, if the directions of the vectors in a region have a higher variation, it is more difficult to predict the trajectory of flow within, hence a higher uncertainty. Based on this idea, we can use Shannon's entropy to measure the complexity of a vector field.

To calculate the entropy value from a collection of vectors, the first step is to compute the corresponding probability distribution function. We can achieve this using the histogram, a discrete representation of the probability distribution function. As we know, a two- or three-dimensional vector has two or three components. If we compute a histogram for each of the components separately, we could not capture the correlation between the directional components, which is undesired. A better solution is to subdivide a unit circle or sphere into a number of segments or patches of equal size, where the number of subdivisions determines the level of discretization.

Figure 5.1 shows several vector histograms displayed as polar plots created from the discretization of two-dimensional vector fields, where each bin represents the frequency of vectors that fall into the segment in the circle. For three-dimensional data, we can perform the subdivision by repeatedly subdividing each triangle in a icosahedron into three triangles of equal size, until a desired discretization level is reached. As we can see, vectors in regions that have higher variations have more non-empty bins, and vice versa, which are reflected in their entropy values. The entropy is calculated as

$$H(X) = - \sum_{x_i \in \mathbb{X}} p(x_i) \log p(x_i), \tag{5.1}$$

where

$$p(x_i) = \frac{C(x_i)}{\sum_{i=1}^{n} C(x_i)} \tag{5.2}$$

$C(x_i)$ is the number of vectors in bin x_i. Using the probabilities calculated here, we can compute the entropy from Equation 5.1. Vector fields with a higher degree of variation in the vector direction will have higher entropy values, and thus are considered to have more information than vector fields where most of the vectors are parallel. Because a region with a higher directional variation has a higher entropy, regions near critical points can be identified based on the entropy values computed from the data in the neighborhood. Although a high variation of vectors is only a necessary condition for the existence of critical points, the difference in the entropy values computed from vector histograms can be used to indicate regions of interest.

5.1.1   Entropy Field

To characterize the complexity of vector data in a local region, given a vector field, we can calculate a scalar value for each grid point and create a derived scalar field called the *entropy field* [146]. The scalar value at each grid point is computed from the Shannon entropy from the histogram in its local neighborhood. The entropy field indicates where in the vector field the complexity is high, which can guide the user to perform further exploration. As shown in a later section, the user can place more streamline seeds in regions that have higher entropy values as a starting point for exploration.

To compute the entropy field, two parameters need to be decided. One is the neighborhood size and the other is the number of bins used in the histograms. Clearly, if the neighborhood size is too small, the number of samples will not be sufficient for constructing the distribution and hence the resulting

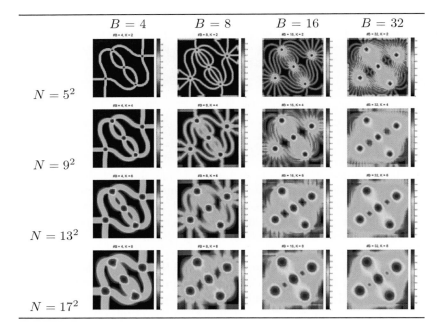

Figure 5.2 The entropy fields computed from the vector field in Figure 5.1(a) using different neighborhood size (N) and numbers of histogram bins (B) [146].

entropy will not be stable. On the other hand, if the neighborhood size is too large, the cost to collect samples in the surrounding region will be higher and the difference in the entropy values in adjacent points will diminish.

Figure 5.2 shows the entropy fields computed from the vector field in Figure 5.1(a) using a different number of bins B and neighborhood sizes N. Higher entropy values are mapped to warmer colors. The neighborhood size is related to the number of samples used to calculate the histogram for entropy computation. An insufficient sample size can create artifacts in the resulting entropy field. The image at $N = 5^2, B = 32$ shows an example of insufficient neighborhood size where the number of samples at each point is smaller than the number of bins. The artifacts can be alleviated by increasing the neighborhood size, as shown in the images where $N = 9^2$, 13^2, and 17^2. For points near the domain boundary where the neighborhood exceeds the boundary, we mirror the data to compute the entropy. The benefit of using entropy fields is that they can highlight not only regions near the critical points but also regions that are more complex.

5.2 COMPLEXITY MEASURES OF STREAMLINES

5.2.1 Information Complexity of Streamlines

Displaying streamlines is one of the most popular methods for visualizing two- and three-dimensional vector fields. How to select streamlines for display such that the user can easily identify the underlying field's flow features, however, is not a trivial problem. When too many streamlines are shown, the resulting visualization can become very cluttered. On the other hand, if not enough streamlines are displayed, important flow features cannot be revealed.

Ideally, since streamlines are used to depict the underlying flow patterns in the field, an ideal set of streamlines should carry the same amount of information as the field. If we represent the data in the vector field as a random variable X, and the displayed streamline set as a random variable Y, the information shared between X and Y can be modeled using the mutual information $I(X;Y)$. If the streamline set Y fully represents the vector field X, the mutual information $I(X;Y)$ should be maximized.

Directly computing the mutual information between a vector field and a set of streamlines is non-trivial. This is because it is difficult to find corresponding samples from the two variables. A typical vector field is often approximated by discrete samples defined at grid points of a mesh. The streamline set, on the other hand, is sparsely distributed over the domain. As a result, to calculate the mutual information between a vector field and a set of streamlines, it is necessary to perform transformations of some sort to match these two different representations.

The goal of displaying streamlines is to allow the user to easily infer the directional information in the vector field. To evaluate whether this goal is achieved, we can reconstruct a new vector field from the given streamlines alone. To do this, a streamline can be thought of as a collection of tangent vectors. These tangent vectors can be used as the known data samples in the field. A discrete vector field can then be reconstructed from these known samples. There exist multiple ways to reconstruct a data field from scattered samples. In a later section, we describe one method to reconstruct the new vector field.

For now, we assume that the reconstructed vector field is already available, which is then used as the proxy to represent the streamline set. With this proxy vector field, we can evaluate how well the streamlines represent the original vector field. As mentioned above, we denote the original vector field as a random variable X, and the streamline-based reconstructed field as a random variable Y. The conditional entropy of X given Y can be computed as

$$H(X|Y) = \sum_{y \in \mathbb{Y}} p(y)H(X|Y = y),\qquad(5.3)$$

where

$$H(X|Y = y) = -\sum_{x \in \mathbb{X}} p(x|y) \log p(x|y).\qquad(5.4)$$

To calculate the conditional entropy between the two vector fields X and Y, we first estimate the joint probability distribution between the two fields by constructing a two dimensional joint histogram, where each vector field is represented by a dimension. We can partition the range of the vector into 60 bins for 2D vector data, and 360 bins for 3D vector data. The histogram cell (i, j) is incremented by one if a vector x in the input vector field is in the range of i-th bin, and a vector y in the reconstructed vector field of the same location is in the range of j-th bin. The joint histogram is an approximation of the joint probability distribution function, which can be used to derive the marginal probability of $p(y)$ and the conditional probability $p(x|y)$ for the equations above.

Two mathematical properties of the conditional entropy are important to our problem. First, the upper bound of $H(X|Y)$ is $H(X)$. This means that given two vector fields, the conditional entropy computed from the original vector field that contains more information will be higher, thus requiring more streamlines. Second, $H(X|Y)$ measures the remaining entropy, or uncertainty, of X given that the values of the random variable Y are known. In the context of our problem, this indicates how much information in the original vector field remains unknown given the reconstructed vector field derived from the known streamlines. The conditional entropy will converge to a small value if enough streamlines have been placed. This property can be utilized to avoid the placement of unnecessary streamlines, therefore reducing visual clutter.

5.2.2 Geometric Complexity of Streamlines

In addition to measuring the complexity of a streamline set through the proxy vector field, we can also measure the geometric complexity of one or multiple streamlines [81]. To do this, we construct the distributions of certain pre-selected measures from the sampling points along the streamline curves. The choice of the measures depends on the underlying application and the user's need. For example, when the user is looking for a specific type of flow feature such as vortices, winding number [92], helicity [39], and Λ_2 [66] can be used. When the user is more concerned with the geometric shape of streamlines, curvatures or torsion can be used since they are considered as the fundamental properties of space curves. Since our focus in this section is the geometry of a streamline, we provide the definitions of curvature, curl, and torsion.

Curvature

The curvature describes the rate of change in the tangent vector over a space curve with respect to the length of the arc. Let T denote the unit tangent vector at a point on a streamline and S denote the length of the arc; the curvature κ is the magnitude of the rate of change of T: $\kappa = \|\frac{dT}{dS}\|$.

Curl

Given a vector field V, the curl is denoted as $\nabla \times V$. The curl is a vector measure of the rotation at a point P in the vector field. The vector itself represents the axis of the rotation, and the magnitude of the vector denotes how fast it rotates. Only the magnitude of the curl is used in our case since we are more interested in the tendency of circulation in a flow field. The curl of a point in vector field V can be defined as

$$\nabla \times V = (\frac{\partial V_z}{\partial y} - \frac{\partial V_y}{\partial z})\mathbf{i} + (\frac{\partial V_x}{\partial z} - \frac{\partial V_z}{\partial x})\mathbf{j} + (\frac{\partial V_y}{\partial x} - \frac{\partial V_x}{\partial y})\mathbf{k}.$$

Torsion

The torsion of a streamline measures the degree of twisting around its binormal vector. It also describes the rate of change of the streamline's osculating plane. If the torsion is zero, it means there is no twisting around the binormal vector, and the streamline completely lies on the osculating plane defined by the tangent and normal vectors. The torsion τ of a streamline is given by $\tau = -N \cdot B'$; $B = T \times N$, where N, T, and B are the normal, tangent, and binormal vectors, respectively, and B' is the derivative of the unit binormal vector.

5.2.2.1 Construction of Histograms

Since the property of a streamline can vary from point to point, it is not sufficient to keep only a 1D histogram because we are not able to preserve position-dependent information such as where in the streamline the curve becomes more winding. To encode this information, we can use a 2D histogram by dividing a streamline into multiple segments and keep a 1D histogram for each segment. Organizing multiple 1D histograms across a streamline serves two purposes. One is to make it easy to identify the most salient portion of

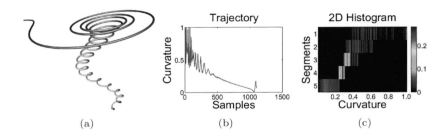

Figure 5.3 (a): The selected streamline color coded by curvature. (b) The curvature values ordered along the streamline. (c) The 2D histogram representation for this streamline; color represents frequency [81].

the streamline, for example, the portion that has the maximum average curvature or the highest entropy in the vector direction. The other is to allow us to compare streamlines. The use of 2D histograms is a compromise between storing the streamline's geometric property at every sample point in a sequence, where the positional information is fully preserved, and storing the histogram across the entire streamline, where no positional information is kept. At one extreme, when every sample point on a streamline represents a segment, the 2D histogram becomes a point-based metric. At the other extreme, if the entire streamline is treated as a single segment, only an 1D histogram is kept. The issue of how many segments to keep for a streamline depends on the complexity of the streamline, and is discussed later.

An issue related to histogram construction is to decide the optimal number of bins (or the bin width) of each 1D histogram. In general, determining the optimal number of bins is non-trivial, and can be computationally expensive. There exist several empirical rules for this purpose. One example is to use Scott's choice [104] to decide the bin width for the 1D histogram in each streamline segment. Scott's choice decides the bin width based on the sample's standard deviation and the number of samples: $K = \frac{3.5\sigma}{N^{1/3}}$, where K is bin width, σ is the sample standard deviation, and N is the number of samples.

An example of a 2D histogram constructed in this way is shown in Figure 5.3. Figure 5.3(c) shows the 2D histogram of the curvature for the streamline shown in Figure 5.3(a). Based on the 2D histogram representation of the streamline, we can understand some underlying features about the streamline. For example, from Figure 5.3(c), the represented streamline has a high curvature with high variation at the beginning, but it gradually decreases to a low curvature. This can be observed from the curvature values ordered along the streamline, as shown in Figure 5.3(b). Since the 2D histogram incorporates the order information to some extent, it can better differentiate streamlines than simply using 1D histograms.

Figure 5.4 A segmentation result where segments are shown by colors [81].

5.2.3 Streamline Segmentation

The basic of idea of dividing a streamline into multiple segments is to ensure that streamline segments of very different complexity are separated to preserve the positional information. In other words, a streamline should be split into different segments if the segments have very different characteristics in their distributions. To segment a streamline, we start with an entire streamline, and then choose a split point where the difference in the distributions of the two segments are the maximum. A split is allowed only when the difference in the distributions of two segments is greater than a threshold. After a streamline is split into two segments, we continue the process recursively and terminate the segmentation process until the difference falls below the given threshold. This streamline segmentation process requires two parameters: one is the minimum length of streamline segments and the other parameter is a threshold ε used to decide whether or not we should split a streamline segment. Figure 5.4 shows a result of segmentation for a streamline based on curvature.

There exist several standard methods to measure the distance between two 1D histograms. Examples are, the L1 distance, Euclidean distance (L2), Kullback–Leibler distance (KL) [74], Bhattacharyya distance [4], and earth mover's distance (EMD) [96]. These distance measures can be classified into two categories: bin-to-bin distance functions and cross-bin distance functions. The bin-to-bin distance functions, such as, the L1 distance, Euclidean distance, Kullback–Leibler distance and Bhattacharyya distance, do not consider the correlation between non-corresponding bins when computing the distance between histograms. On the other hand, the cross-bin distance functions, such as the earth mover's distance, considers the cross-bin relationship, albeit at higher computation cost.

5.3 INFORMATION-AWARE STREAMLINE SEEDING

In this section, we show how information theory can help us analyze vector fields in a more effective way. The key concept used here is to quantify the amount of information that has been extracted by the visualization algorithms, and then adjust the parameters of the visualization algorithms so that hidden information in the dataset can be revealed. We demonstrate the use of information theory in placing streamline seeds, an important process when using streamlines to visualize a vector field. We first describe an object-space approach for streamline seeding without considering the view parameters in 3D visualization. Then we explain how seeding can be further optimized if we exploit view-dependent information to maximize the placement of streamline seeds.

5.3.1 View-Independent Method

5.3.1.1 Initial Seeding

Since particles originated from local regions in a vector field that have higher variations in direction can have very different trajectories after they are traced out, it is necessary to place more seeds there so as not to miss any important flow features. To do this, an entropy field is computed from the vector data as described above, and then the local maximum points are used as the initial seed placement candidates. The local maximum points whose entropy values are too small are discarded, for example, less than 70% of the maximum entropy value in the field. The seeds are distributed using a diamond-shaped template, which can be seen from the location of the red points in Figure 5.5(b) and 5.5(h). The template used here is inspired by the flow topology-based method proposed in [135]. Since we only choose regions that have larger entropy values without explicitly detecting the existence of critical points and their types, the template is a combination of the templates used in [135], whose aim is to capture the local flow topology near the salient regions. For 3D data, an octahedral-shaped template is used. Each template places 27 seeds: one seed is placed at the centroid of the octahedron; 6 seeds are placed at the 6 vertices; 8 seeds are placed at the centers of the 8 faces, and 12 seeds are put at the midpoints of the 12 edges.

5.3.1.2 Importance-Based Seed Insertion

Although we placed streamline seeds around regions that have high local entropy, there is no guarantee that the entire field will be properly covered by the resulting streamlines. Typically, there exist regions that have too many streamlines, and there are void regions whose information remains hidden. To evaluate how well the current set of streamlines represent the entire vector field, we reconstruct a vector field from the streamlines, and compute the conditional entropy between the original field and the reconstructed proxy field as

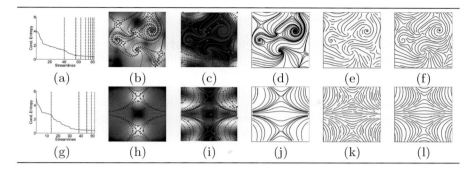

Figure 5.5 The process of streamline generation for two different 2D vector fields (top and bottom) [146]. (a) and (g) plot the values of the conditional entropy as more streamlines are placed in the fields. The vertical black dashed lines indicate the end of each iteration. (b) and (h) show the initial seeds (in red) and the resulting streamlines (in blue) using our seeding template. (c) and (i) show the resulting streamlines after the first iteration of importance-based seeding. (d) and (j) show the streamlines when the conditional entropy has converged. (e) and (k) are the streamlines generated by the evenly spaced seeding method [69]. (f) and (l) show streamlines generated by the farthest-point seeding method [84].

described previously. We model the placement of additional streamline seeds as an importance-based sampling problem where the probability to place a seed at a location is proportional to the conditional entropy computed from the point's local neighborhood. Given the conditional entropy at each point, the expected probability $p(x, y)$ of placing a seed at (x, y) is computed by the following equation:

$$p(x, y) = \frac{h(x, y)}{\sum_{\forall x, y} h(x, y)}, \tag{5.5}$$

where $h(x, y)$ is the conditional entropy at point (x, y). The probability at each point forms the probability distribution function (PDF) of seeding in the domain. With the seeding PDF, we can distribute the seeds according to the probability at each point by importance sampling. As a well-studied problem in statistics, importance sampling can be achieved by either the inverse transformation method or the acceptance-rejection method. In this framework, we use the inverse transform method with the chain rule for multidimensional data.

5.3.1.3   Redundant Streamline Pruning

Even with our importance-based sampling method, it is still possible that not all the streamlines contribute to the understanding of the data. To evaluate whether a streamline is redundant or not, one can compute the difference in the conditional entropy with and without the streamline. Unfortunately, this is too expensive because we will have to repeatedly update the reconstructed vector field when evaluating the existing streamlines one by one.

To speed up the process, we can use an information-assisted distance-based approach. Intuitively, if a streamline is close to any of the existing streamlines, there is a possibility that it contributes very little to the understanding of the underlying vector field. In other words, in a low-entropy region, where fewer streamlines are needed, the possibility for a streamline to be redundant is higher compared to the case when the streamline is in a higher-entropy region. Based on this idea, we determine whether a streamline should be pruned based on whether there are any other streamlines in the local region within a distance threshold R. To account for the entropy values when pruning redundant streamlines, we allow higher-entropy regions to have a smaller distance threshold than lower-entropy regions. We choose a distance threshold from a scalar range $[R1, R2], R1 < R2$, where $R1$ is used for regions that have the maximum entropy and $R2$ is used for regions with the minimum entropy. Regions that have entropy in between will use a threshold linearly interpolated between $R1$ and $R2$. In our experiment, we set $R2$ to be 2% of the smallest dimension in the field, and $R1$ at half of $R2$.

It is noteworthy that the result of pruning is order dependent. That is, if we place less important streamlines first, more salient streamlines will be pruned if they are too close to the unimportant ones. To solve this problem, the streamlines are first sorted based on the entropy values at their seeds in a decreasing order before the pruning process starts. Therefore, more salient streamlines will have a higher chance to survive.

The above importance-based sampling and seed pruning steps are performed iteratively. Within each iteration, a pre-determined number of seeds will be introduced to the field. We set the number of seeds in each iteration as the square root of the number of grid points. The process is repeated until the conditional entropy between the original vector field and the proxy vector field converges.

5.3.1.4   Reconstructing a Vector Field from Streamlines

To create the proxy vector field from a given set of streamlines, a straightforward method is to use a low-pass filter such as the Gaussian kernel to smooth out the vectors on the streamlines to their surrounding regions. This method, however, has several drawbacks. First, choosing an appropriate kernel size for the filter is non-trivial. Kernels that are too small will not be able to cover the entire field when only a sparse set of streamlines are present. On the other hand, when the size of the kernel is too large, important flow features in the field may be destroyed.

We formulate the problem of generating the proxy vector field as an optimization problem that tries to generate a vector field $\mathbf{Y}(\mathbf{x} = (x, y, z)) = (u(\mathbf{x}), v(\mathbf{x}), w(\mathbf{x}))$ with respect to the field $\hat{\mathbf{X}}(\mathbf{x})$ that minimizes the following energy function:

$$\varepsilon(\mathbf{Y}) = \int \varepsilon_1(\mathbf{Y}(\mathbf{x}), \hat{\mathbf{X}}(\mathbf{x})) + \mu\varepsilon_2(\mathbf{Y}(\mathbf{x}))d\mathbf{x}, \tag{5.6}$$

where

$$\varepsilon_1(\mathbf{Y}(\mathbf{x}), \hat{\mathbf{X}}(\mathbf{x})) = |\hat{\mathbf{X}}(\mathbf{x})|^2 |\mathbf{Y}(\mathbf{x}) - \hat{\mathbf{X}}(\mathbf{x})|^2$$
$$\varepsilon_2(\mathbf{Y} = (u(\mathbf{x}), v(\mathbf{x}), w(\mathbf{x}))) = |\nabla u(\mathbf{x})|^2 + |\nabla v(\mathbf{x})|^2 + |\nabla w(\mathbf{x})|^2.$$

In the term $\varepsilon_1(\mathbf{Y}(\mathbf{x}), \hat{\mathbf{X}}(\mathbf{x}))$, $\hat{\mathbf{X}}(\mathbf{x})$ equals the original field $\mathbf{X}(\mathbf{x})$ along the streamlines and zero elsewhere. Therefore, ε_1 is always zero regardless of the intermediate vector $\mathbf{Y}(\mathbf{x})$ for $\mathbf{x}$ not occupied by the streamlines since $\hat{\mathbf{X}}(\mathbf{x})$ is zero. This term will be minimized when the reconstructed vector $\mathbf{Y}(\mathbf{x})$ along the streamlines is equal to the original vector $\mathbf{X}(\mathbf{x})$. The term $\varepsilon_2(\mathbf{Y})$ is the sum of the gradient magnitude of the three components u, v, and w at location $\mathbf{x}$, which can be minimized when the neighboring vectors are identical.

In other words, the diffusion process is modeled as a constraint optimization problem with a soft boundary condition, where ε_1 penalizes the violation of the boundary condition along the streamlines, and ε_2 measures the smoothness of the reconstructed vector field, where the trade-off between the boundary condition and the smoothness is controlled by the parameter μ. By setting $\mu = 0.1$ in our experiments, empirically, the reconstructed vector converges to the input vector along the streamlines, thus preserving the boundary condition.

It is noteworthy that the energy equation presented here is similar to the force used in the *gradient vector field Snake* [145], which can be solved by the *generalized diffusion equations* described in the fluid flow literature [21]. More details about the solution and the convergence of the diffusion can be found in [145]. In the image processing field, there exists similar research such as [67, 90] that distributes colors on a curve to its surroundings. The major difference between the two is that the gradient constraint in the diffusion curve method demands high gradients on the input curves to preserve object boundaries. In streamline diffusion, the smoothness constraint is applied everywhere, even for the streamline boundaries. The intermediate field computed by streamline diffusion is used to measure the streamline information and the placement of seeds as explained in the following sections.

5.3.1.5   Seed Selection Result

Figure 5.6 shows streamlines generated from a 3D dataset *Plume*, which simulates the thermal downflow plumes on the surface layer of the sun by scientists

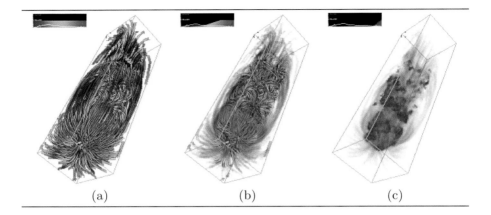

Figure 5.6 Streamline rendering results for the dataset *Plume* [146]. In (a), salient streamlines (in warm color) are occluded. In (b) and (c), the occlusion is reduced by modulating the opacity according to the entropy values. The transfer functions used to map the entropy to color/opacity are shown at the upper left corners.

at the National Center for Atmospheric Research. The resolution of the data is $126 \times 126 \times 512$. In Figure 5.6(a), the first 600 streamlines generated by our method are shown. The color near the internal region is warmer, indicating higher entropy values and hence more complex flow in that region. In this image, the detailed structure of these warmer-colored streamlines are occluded by the blue streamlines with smaller entropy, causing difficulties for more detailed analysis. To reduce the occlusion, the entropy field can be utilized to adjust the color and opacity of the streamline segments and make those streamlines in regions of lower-entropy less opaque. Figures 5.6(b) and (c) show two rendered images using different transfer functions shown at the upper left corner of each sub-figure. Figure 5.6(c) shows a result that maps the low-entropy region to a smaller opacity, revealing a clearer structure in the high-entropy region, while Figure 5.6(b) provides more cues about the streamlines in the outer regions as a context.

5.3.2 View-Dependent Method

5.3.2.1 Maximal Entropy Projection (MEP)

One particular challenge for effectively visualizing a dense set of three-dimensional streamlines is to overcome the occlusion problem. Placing more streamlines near regions of higher complexity in three-dimensional space does not always guarantee a clear view of salient flow features. In this section, we show how this problem can be overcome by a view-dependent streamline

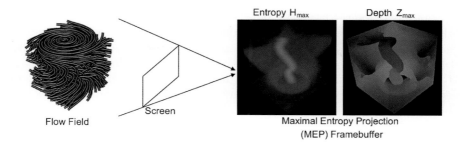

Figure 5.7 Illustration of the MEP-framebuffer. A MEP framebuffer consists of two buffers H_{max} and Z_{max} that, for a given viewpoint, store the maximal entropies for all pixels and the corresponding depths of the voxels with maximal entropy [76].

placement method, which can capture more complex flow regions in image space.

The view-dependent algorithm starts with an entropy field defined in 3D object space, as described above. To locate which region in the screen potentially has more salient flow features, we project the entropy value from object space to image space according to the camera view, and for each pixel, we keep the maximum entropy value, referred to as the maximum entropy projection (MEP). This is similar to the conventional maximum intensity projection (MIP) technique used in medical imaging. This maximum entropy value at each pixel is stored in a MEP entropy buffer, and the corresponding z value is stored in a MEP z-buffer. Together these two buffers are called a MEP framebuffer. With the MEP framebuffer, we know where the more complex flow features are located in the screen, and their distances to the image plane, based on which we control the placement of streamlines to minimize occlusion. Figure 5.7 illustrates an example of a MEP framebuffer [76].

5.3.2.2 Seed Placement

Given a density of streamlines in object space that have been computed, based on the MEP framebuffer, the goal is to select the more salient streamlines to display for the given view. This can be done by assigning each streamline a priority score ω based on their screen projection, which will be used as the priority score of the streamline. A streamline with a higher-priority will contribute to revealing more complex flow features and a lower-priority streamline will have very little benefit or even cause occlusion.

To compute the priority score ω for a streamline, we project the streamline to the image plane and give a score ω_f to each fragment of the streamline. To compute ω_f, the basic idea is that when a fragment is in front of the point that has the maximum entropy, the priority score will decrease if the distance between the fragment and the maximum entropy point increases, and the difference between the entropy values of the fragment and the maximum entropy becomes larger. This is because the fragment is not near the region of interest (high flow entropy) but will cause occlusion. On the other hand, if the fragment is behind the point that has the maximum entropy, although it does not cause any occlusion, its priority score will decrease as the fragment becomes farther away from the maximum entropy point. Specifically, if z_w is smaller than $Z_{max}(x_w, y_w)$, the value of ω_f is computed as ω_{front} in Equation 5.9:

$$\triangle H = H_{max}(x_w, y_w) - H(x_o, y_o, z_o) \tag{5.7}$$

$$\triangle Z = z_w - Z_{max}(x_w, y_w) \tag{5.8}$$

$$\omega_f = \omega_{front} = -H_{max}(x_w, y_w)(1 - e^{-|\triangle H||\triangle Z|}), \tag{5.9}$$

where $H_{max}(x_w, y_w)$ is the maximum entropy along the current view ray, and $\triangle H$ and $\triangle Z$, respectively, are the difference of entropy and depth from the this fragment to the values in the corresponding pixel in the MEP framebuffer. Since this fragment can cause occlusion, its weight is negative so it can reduce the score of this streamline, while the amount of reduction in the score depends on the difference of depth and the entropy. If the depth or entropy of this fragment are near those in the MEP framebuffer, it is near the salient flow feature that is visible from this pixel and thus the reduction should be small. We design this equation such that the reduction is zero when $\triangle H$ or $\triangle Z$ is zero, and close to $H_{max}(x_w, y_w)$ when $|\triangle H|$ and $|\triangle Z|$ is very large.

For the fragments that are behind the maximum entropy point, i.e., the depth of the fragment z_w is larger than or equal to $Z_{max}(x_w, y_w)$, and since they do not cause occlusion, the score ω_{back} is computed as in Equation 5.10:

$$\omega_f = \omega_{back} = H(x_o, y_o, z_o)e^{-|\triangle H||\triangle Z|}, \tag{5.10}$$

where $\triangle H$ are $\triangle Z$ are defined in Equations 5.7 and 5.8, respectively. While ω_{back} is always positive, it becomes smaller as the fragment moves farther away from the point of maximal entropy, or as the difference between the entropy values becomes larger. Meanwhile, similar to ω_{front}, ω_{back} considers the entropy of the fragment and receives a lower score if the entropy is low.

With ω_f computed for each fragment of a streamline, the sum of the scores from all fragments are summed together as the priority score of the streamline:

$$\omega_s = \sum_{f \in s} \omega_f. \tag{5.11}$$

The priority score computed for each streamline can be positive or negative, where larger values indicate more salient streamlines. Figure 5.8 shows the result of streamlines evaluation for the dataset *Tornado*. It can be seen

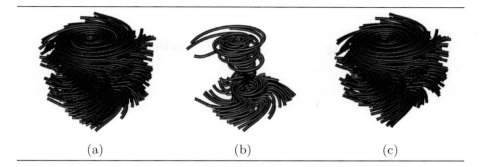

(a) (b) (c)

Figure 5.8 View-dependent streamline evaluation for *Tornado*. (a) The initial set of streamlines where the blue and red ones have positive and negative scores, respectively. (b) The streamlines with positive scores. (c) The streamlines with negative scores [76].

that the streamlines with positive scores (in red) are around the center of the tornado or at the bottom of the flow since they do not cause much occlusion, while streamlines with negative scores (blue) can cause occlusion and hence should have a lower priority for display.

5.3.2.3 Entropy-Based Streamline Selection

The priority scores computed for the streamlines for a given view allow us to discard unimportant streamlines or streamlines that will block important flow features. To choose salient streamlines, one simple way is through thresholding, that is, streamlines that have priorities lower than the threshold are removed. However, sometimes it is difficult for the user to choose a proper threshold. To automate the streamline selection process, we can make the density of streamlines in any given screen area proportional to the complexity of the flow. To do this, for a given screen area, we can estimate the streamline density in a local region in screen space by dividing the number of pixels occupied by the streamlines by the total area of the region in pixels. We can also compute for each region on the screen an expected streamline density, which is defined as the average normalized entropy of the region in the MEP framebuffer. A region is too crowded if the streamline density exceeds the expected streamline density. To decide whether a new streamline should be selected, the local neighborhood around the pixels along the streamline is checked. A streamline is placed only if the local neighborhood is not overly crowded. It is worth mentioning that this streamline selection algorithm may generate different results if the streamlines are processed in different orders. Streamlines that are checked earlier will have a better chance to be selected. Therefore, to place streamlines that can highlight more salient regions with less occlusion, streamlines with higher priority scores should be checked first. Otherwise, streamlines that are selected earlier can preclude streamlines checked later even if they have higher priority scores.

5.3.2.4 Finding Optimal Views

With the entropy field and MEP framebuffer, it is possible to select optimal views to visualize the pre-generated streamlines. When the user tries to find the best viewpoint, intuitively s/he would like to see as many interesting features of the flow, preferably with minimum occlusion.

Given a set of streamlines, the optimality of a viewpoint can be evaluated based on the complexity of the projected flow on the screen. In the MEP framebuffer, each pixel records the maximum entropy encountered by a viewing ray. The sum of the entropy values in the MEP framebuffer can provide an upper bound of the complexity, and the mean entropy can provide an average complexity visible to the viewer. Since our goal is to maximize the flow complexity visible to the viewer after projection, we choose the sum of the entropy values as the priority score for a viewpoint.

To search for the optimal viewpoint, we first construct a domain of candidate viewpoints, which is a sphere centered at the center of the flow field. This sphere is then tessellated into uniform triangles, where each viewpoint is placed at the center of a triangle. Once the entropy sum in the MEP framebuffer for each viewpoint has been computed, the viewpoint with the highest score is then selected as the optimal viewpoint.

Figure 5.9 shows examples of optimal viewpoint selection. The leftmost column in Figure 5.9, displays the scores of the viewpoints for the dataset *Tornado*, where blue to red represents scores from low to high. We can see that more than one viewpoint receives high scores. In the three examples presented here, the center of the sphere in each of the images is used as the focus point to generate the corresponding images. From the images, it can be seen that the scores of the viewpoints change smoothly, which is consistent with our expectation of spatial coherence.

Visualizations of better and worse viewpoints for the dataset *Tornado* [34] are shown in Figure 5.9. As mentioned earlier, better viewpoints reveal more information about the flow, which implies that the image should have less occlusion and can highlight the shapes of the streamlines better. From the images in the top two rows of Figure 5.9, it can be seen that there exist two types of less desired viewpoints for this dataset. One type of viewpoint displays the data from the side, as shown in Figure 5.9 (a). In Figure 5.9(a.1), a blue stripe surrounds the tornado, where the corresponding viewpoints do not allow the viewer to see the streamlines of high curvature. The other type of undesired viewpoint is the one that views the streamlines from the top, as shown in Figure 5.9(b.2). Although from this viewpoint one can visualize the streamlines with high curvature, the depth cue is lost. By comparing with Figure 5.9(a.3) and (c.3), the MEP framebuffer in (b.3) shows that the region of high entropy is self-occluded. The bottom row of the figure represents a good viewpoint for this dataset. Compared to the less desired viewpoints, Figure 5.9(c.2) shows the viewpoint with the highest score, which can display not only the streamlines with high curvatures but also display the 3D nature of the dataset quite well.

The rightmost column in Figure 5.9 shows the streamlines selected by the streamline selection algorithm for the corresponding viewpoints. While visual cluttering and occlusion are reduced compared to using all streamlines as shown in the second column, because of the use of non-optimal viewpoints, the results in Figure 5.9(a.4) and (b.4) are not as good as in Figure 5.9(c.4).

5.4   INFORMATION CHANNELS FOR FLOW VISUALIZATION

Another approach for streamline and viewpoint selection in flow visualization are discussed in Tao et al. [119] by setting an information channel between the set of viewpoints $V = \{v_1, v_2, ..., v_m\}$ and the set of streamlines $S = \{s_1, s_2, ..., s_n\}$. The assumptions for viewpoints are that the flow field is centered in a sphere of sample viewpoints and that the camera at a sample viewpoint is looking at the center of the sphere. As in the visibility channel defined in Section 3.3 (see Figure 3.7), the main components in channel $V \to S$ are:

- The transition probability matrix $p(S|V)$ where conditional probabilities $p(s|v)$ represent the probability of seeing streamline s from viewpoint v (i.e., the importance of s with respect to v).

- The input probability distribution $p(V)$ where $p(v)$ represents the probability of selecting viewpoint v. Assuming $p(v)$ is evenly distributed, then $p(v) = 1/m$ where m is the number of sample viewpoints.

- The output probability distribution $p(S)$ where $p(s)$ represents the average probability that streamline s is seen from all viewpoints V. That is, $p(s) = \sum_{v \in V} p(v)p(s|v)$.

Similar to the inversion of the visibility channel 3.3.2 (see Figure 3.8), an inverted information channel $S \to V$ can be defined, where the input and output probability distributions are swapped: $p(S)$ becomes the input and $p(V)$ becomes the output. In this inverted channel, the new transition probability matrix is $p(V|S)$, where $p(v|s)$ represents the probability of selecting viewpoint v given streamline s. The probabilities $p(v|s)$ are obtained from $p(s|v)$ using Bayes' theorem.

Observe that to construct the channel we need to know the conditional probabilities $p(s|v)$. To obtain them, [119] proceeds as follows. Each streamline s and its projection from viewpoint v, s_v, are considered as random variables built from a finite set of points, and their mutual information is obtained:

$$I(s; s_v) = \sum_{i \in s} \sum_{j \in s_v} p(i, j) \log \frac{p(i, j)}{p(i)p(j)}. \tag{5.12}$$

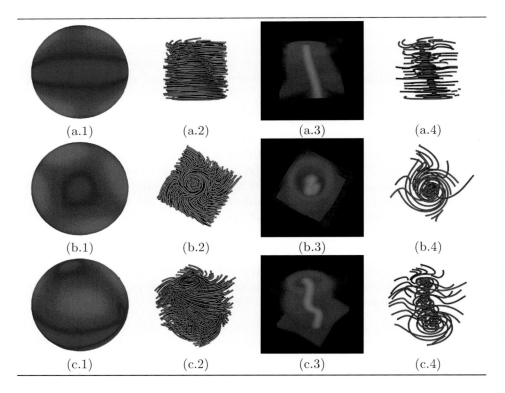

Figure 5.9 Viewpoint selection for *Tornado*. From left to right, scores for the corresponding viewpoints, the vector field, the MEP frame-buffer, and the result from the view-dependent streamline placement algorithm. In the leftmost column, the center of the sphere's projection encodes the score for the corresponding viewpoint. Scores are mapped from blue (low) to red (high). Figures in the top two rows show two poor viewpoints. The top row shows one of the low-score views. The tornado is viewed from one of the sides and thus the circular pattern of the flow field is not revealed. Bottom row: one of the good viewpoints [76].

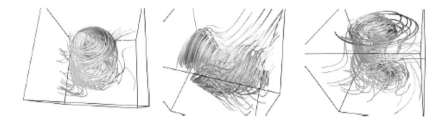

Figure 5.10 View selection for flows [119]: The best viewpoint selection obtained with mutual information only (left), shape characteristics only (middle), and both mutual information and shape characteristics (right).

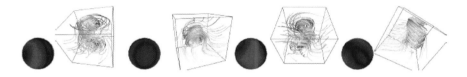

Figure 5.11 Tornado dataset [119]: In the view sphere images, red to blue is for best viewpoint to worst viewpoint. Streamline rendering from the best viewpoint and the worst viewpoint is also shown.

To compute $p(i)$, vectors from the original flow data are interpolated based on the positions of all the points along s. These vectors are used to construct a 2D histogram based on vector magnitude and (quantized) direction. To compute $p(j)$, the projections of these vectors along s_v are used to construct the corresponding histogram. The joint probability $p(i, j)$ is computed from the joint histogram for s and s_v where each of the two axes consists of all vector directions and magnitude combinations.

The mutual information $I(s; s_v)$ is then weighted by the shape characteristics of the streamline skeleton, $\tilde{\alpha}(s; v)$ which indicates how stereoscopic the shape of streamline s is reflected under viewpoint v, and $p(s|v)$ is then defined as

$$p(s|v) = \frac{\tilde{\alpha}(s; v)I(s; s_v)}{\sum_{s \in \mathbb{S}} \tilde{\alpha}(s; v)I(s; s_v)}. \qquad (5.13)$$

Figure 5.10 shows the difference between using or not using the streamline shape in the definition of $p(s|v)$.

From the channel $S \rightarrow V$, the best streamlines are selected, either by using the maximum value of $p(s)$ or the minimum value of streamline information,

Figure 5.12 Streamline selection [119]. From top to bottom: hurricane, car flow, computer room, and solar plume datasets. From left to right: initial set, and best selections based on $p(s)$ and $I(s; V)$, representative (REP) selection, and random selection.

defined by

$$I(s; V) = \sum_{v \in \mathbb{V}} p(v|s) \log \frac{p(v|s)}{p(v)}. \tag{5.14}$$

From the channel $V \to S$ the best viewpoints are selected, either by using the maximum value of $p(v)$ or the minimum value of viewpoint information, defined by

$$I(v; S) = \sum_{s \in \mathbb{S}} p(s|v) \log \frac{p(s|v)}{p(s)}. \tag{5.15}$$

Streamline clustering and viewpoint partitioning can also be obtained from both $S \to V$ and $V \to S$ channels, respectively, by first selecting representative streamlines (viewpoints respectively); these representative streamlines (respectively viewpoints) define the clusters and then the other streamlines (respectively viewpoints) are included in the cluster such that the decrease of mutual information of the channel is minimum (Equation 1.34).

5.5 SUMMARY

This chapter introduced several information-theoretic metrics for analyzing the complexity of vector data. We first described the concept of the entropy field, which is used to measure the complexity of field data. The entropy field can indicate the uncertainty of vector directions in the local neighborhood, which is strongly related to the existence of critical points or turbulent flows. We also described two methods to measure the complexity of flowlines: one is based on the mutual information and the other is based on attribute statistics sampled from the flowlines. With the complexity measures, we introduced several information-aware streamline seeding algorithms in both object and image space. The object-space approach uses an energy minimization method to avoid creating redundant streamlines, and the image space method minimizes occlusion and finds good viewpoints. Finally, we discussed how the concept of information channels can be used to model the relationship between viewpoints and streamlines and optimize the rendering images.

FURTHER READING

Kullback, S. (1997). *Information Theory and Statistics (Dover Books on Mathematics)*, Dover Publications, Mineola, NY

Laramee, R.S., Hauser, H., Doleisch, H., Vrolijk, B., Post, F.H., and Weiskopf, D. (2004). The state of the art in flow visualization: Dense and texture-based techniques, *Computer Graphics Forum*, 23(2):203–221.

Lee, T.-Y., Mishchenko, O., Shen, H.-W., and Crawfis, R. (2011). Viewpoint evaluation and streamline filtering for flow visualization, *IEEE Pacific Visualization Symposium*, 83–90.

Lu, K., Chaudhuri, A., Lee, T.-Y., Shen, H.-W., Wong, P.C. (2013). Exploring vector fields with distribution-based streamline analysis, *IEEE Pacific Visualization Symposium*, 257–264.

Tao, J., Ma, J., Wang, C., and Shene, C.-K. (2013). A unified approach to streamline selection and viewpoint selection for 3D flow visualization, *IEEE Transactions on Visualization and Computer Graphics*, 19(3): 393–406.

Xu, L., Lee, T.-Y., and Shen, H.-W. (2010). An information-theoretic framework for flow visualization, *IEEE Transactions on Visualization and Computer Graphics*, 16(6): 1216–1224.

Information Visualization

CONTENTS

Over the last decades, the amount of available data in several fields, such economics, geosciences, or biomedicine, has dramatically increased. This large amount of information has led to the development of several computational methods to derive insights from these data. Among these methods, visualization has acquired high relevance due to the human ability to detect patterns and trends from data visualizations. In the previous chapters, the visualization techniques were focused on data associated with a spatial position. This field is known in the research community as *scientific visualization*. In this chapter, we will focus on data that do not have an assigned spatial position. This field is known as *information visualization*.

Information visualization can have two different objectives that must be taken into account when deciding the visualization design space [17]. First, the main goal can be to communicate or record a certain event or process. In this case, the user already understands the information. Second, the goal can be to analyze the data. In this case, the user wants to get new insights on the data involved in the visualization process.

Information visualization starts with data, which are usually represented as arrays of cases based on variables. In certain situations, a pre-processing step is required to obtain the data in this array form. For instance, when the data to be analyzed are textual, the computation of word occurrence vectors is required. There are three main types of data:

- **Categorical or nominal**: The data can only take values on a limited set, whose elements are usually designated by a name. Comparisons between elements of this type can only be done in terms of "is equal to" or "is not equal to."

- **Ordered**: The data take values on a limited ordered set. Comparisons between elements of this type can also be done in terms of "is greater than" or "is less than" in addition to the equality condition.

- **Continuous**: The data take values in a continuous range. In this case, arithmetic operations are also permitted.

These data values are usually represented in the screen using graphical properties. An elementary visual presentation consists of a set of marks (which could be points, lines, areas, surfaces, or volumes), a position in space-time (the X-Y plane in classical graphics, but 3D space plus time in modern information visualization), and a set of "retinal" properties, such as color, size, or texture [17]. A key feature of these visualizations is that they use the mappings between data and visual vocabulary to provide interactive access to the data.

In this chapter, we will focus on information-theoretic techniques applied to the information visualization process. First, in Section 6.1, we focus on how information theory can be used in order to build theoretical foundations for the information visualization process. In Section 6.2, we show how information-theoretic measures can be used to quantify the visualization quality, focusing in more detail, in Section 6.3, on parallel coordinates visualization. In Section 6.4, we describe the construction of maximum entropy summary trees, which use entropy to decide which information is more relevant to be shown on a tree data structure. In Section 6.5, we present some techniques to get insights on the relationship between different variables in multivariate datasets. In Section 6.6, we go into detail about the analysis of time-varying multivariate data using the transfer entropy measure. In Section 6.7, we explain the use of information-theoretic measures when dealing with sensitive data where the privacy of individual records must be maintained. Finally, in Section 6.8, we introduce mutual information diagrams, which can be used to analyze relationships between different variables in a dataset, and, in particular, how several theoretical models of a process fit the observed data.

6.1 THEORETICAL FOUNDATIONS OF INFORMATION VISUALIZATION

The modern study of information visualization started with computer graphics, but it does not started from scratch, as it adapted a great tradition on the visual representation of data previous to the computer technology. The knowledge of this tradition was based more on qualitative facts than on quantitative measurements. This fact makes this field closer to art than to science. The absence of a theoretical framework for information visualization makes the significance of achievements in this area difficult to describe, validate, and defend [93].

In order to overcome this limitation, several attempts at developing theoretical foundations of information visualization have appeared. The information visualization process is related to data statistics, computer graphics, and perception, among other fields. Due to the complexity of this process, there is no single theory that can include the whole process, and probably, multiple theories at different levels would be needed [93].

Some works [25, 93] have defined information visualization as a communication channel from a dataset to the cognitive processing centered in the human observer. Shannon's theory of information is based on syntactic information (i.e., related to the symbols from which messages are built and to their interrelations), and not on semantic information (i.e., related to the meaning of messages and their referential aspect) or pragmatic information (i.e., related to the usage and effect of the messages). Thus, information theory is probably not capable of effectively explaining certain perceptual aspects of the visualization process, although it has been demonstrated to be a very effective theory to explain the fundamental limits of data communication and compression. From this perspective, an information-theoretic analysis of the visualization process could be an effective way to evaluate the efficiency of the process.

Chapter 2 presents a theoretical framework of visualization based on information theory. This framework is wide enough to comprise scientific visualization as well as information visualization. From this theory, fundamental measures, such as entropy or mutual information, arise to assess the visualization process and quantify its basic properties.

The next sections present some information-theoretic methods and measures that have already been used on the visualization field. Some of them are a consequence of the theoretical framework presented in Chapter 2, while others apply information-theoretic measures in an unrelated way. There is still a need to develop new functionalities and applications based on this theoretical framework. In our opinion, this will be one of the major areas of research in the field of information visualization in the future.

6.2 QUALITY METRICS FOR DATA VISUALIZATION

The most common way to evaluate the performance of a new information visualization technique is through a user study. This type of study has several advantages such as the possibility to analyze different configurations of the technique or to consider high-level perceptual features, but it is not possible to conduct a user study for each individual visualization every time it is created. The quality of visualization images depends on several factors, including the domain-specific requirements, the user's needs and expectations, the source dataset, and the techniques used. Hence, it is desirable to provide users with alternative means for measuring visualization quality. Ideally, such a measure is generic, readily available, and easily applicable to many types of visualization [63].

One of the first attempts to introduce information theory to evaluate visualization performance was proposed by Yang-Peláez and Flowers [147]. In this proposal, the authors defined four types of information content measures: the amount of information spanned by the data, the amount of information spanned by the display, the amount of information in a particular data display, and the amount of topological information content.

The first one, the amount of information spanned by the data, was defined by the authors as a sum of the individual contributions of each data dimension. The authors suggest computing the information content of each dimension as

$$I = \log_2(\frac{range}{precision}). \tag{6.1}$$

Although *range* and *precision* are not well defined for categorical or discrete data values, the idea of this measure is similar to Shannon's entropy considering equiprobable states. Note that with this definition, contrary to Shannon's entropy, the probabilities of each data value are not taken into account. Chen and Jänicke [25] overcome this limitation by introducing a probability mass function of a variable that can be used to describe the probabilistic attributes of a variety of entities, phenomena, and events, such as all instances of data at a sample point, all datasets in a data space, and all visual representations used in an application.

The second type of information refers to the overall data display. Usually, not all the information can be rendered in a single display. In this case, a layout made by multiple displays is used to show different key attributes of the data, and the authors suggest adding the individual display information to estimate the total information.

Third, the maximum information that can be contained by the display is defined as the capacity of the display. For instance, in a $N \times N$ connectivity matrix, where the nodes can be connected or not without any weight, each position of the matrix gives one bit of information, giving total information of $N^2 - N$ bits (if the diagonal positions are omitted).

Finally, the fourth type of information is based on the topological information content, which was defined by Rashevsky in the field of biophysical mathematics [94]. This definition allows us to compute the information of a topological graph for different orders (see [147] for more details).

This first attempt at describing information visualization as an information process has some drawbacks such as its absolute independence of the data to be shown or its marginal use in assessing whether or not a visualization is useful. Later, Chen and Jänicke [25] introduced the theoretical framework presented in Chapter 2 that overcomes these limitations. From this framework, several measures, such as visual information, display space capacity, and potential information loss, have been defined in order to quantify the quality of the visualization from different perspectives. We refer the reader to Chapter 2 for a detailed description of this framework and its related measures.

6.3   IT METRICS FOR PARALLEL COORDINATES

In the visualization pipeline models, the visual representation is usually regarded as the end product, even though it is obviously a central part of the process of visualization and visual analysis. A good visualization has to provide a clear picture of the relevant structures to the user. Optimizing a visualization must not only take the data into account, but has to be done differently for different visualization techniques. Structures that are clearly visible in one technique may be hard to find or cause clutter in another. For the parallel coordinates visualization technique, a quality evaluation framework was presented in [38]. Parallel coordinates is a visualization technique that places in parallel the axis of the different data variables, instead of the orthogonal way in the Cartesian representation [61]. This permits us to easily represent high-dimensional data. In this representation, every element of the data corresponds to a polyline instead of a point, as in the Cartesian representation. Figure 6.1 shows an example of the visualization of the cars dataset [126] using parallel coordinates.

In the above-mentioned work, Dasgupta and Kosara [38] proposed Pargnostics (*Par*allel coordinates dia*gnostics*) as a framework to quantitatively evaluate the visualization technique quality in the whole visualization pipeline. Among other measures, some quality metrics based on information theory were proposed. Parallel coordinates is adequate to show patterns between variables represented in consecutive axes, while the patterns between variables shown in separated axes are not so easily identifiable. Thus, a key point in the visualization design is the axis ordering, typically assigning highly related variables in neighbouring positions. The Pearson correlation coefficient is extensively used for this purpose, but a lack of such correlation does not imply that two variables are independent, since it assumes a linear relationship.

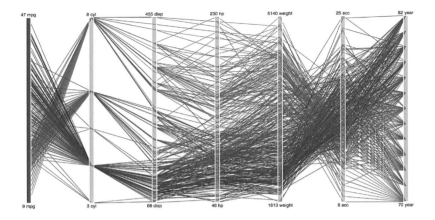

Figure 6.1 Example of the cars dataset visualization using parallel coordinates.

On the contrary, mutual information provides a general measure of dependency between variables. In Pargnostics, the authors model the data dimensions as random variables, and the probability density function estimation is based on the screen space, where each pixel of the screen is considered as a bin of the random variable. The probability that a dimension takes a particular value is equivalent to the probability of a binned value in the screen space. Thus, the marginal and joint probabilities are respectively given by

$$p(x_i) \quad = \quad \frac{b_i}{L}, \tag{6.2}$$

where b_i represents the number of lines that start (or end) at pixel i on the axis of the variable X (or Y) and L is the total number of lines, and

$$p(x_i, y_j) \quad = \quad \frac{b_{ij}}{L}, \tag{6.3}$$

where b_{ij} represents the number of lines that start at pixel i on the axis of the variable X and end at the pixel j on the axis of the variable Y. In Figure 6.2, the construction of the joint histogram required to compute mutual information is illustrated. Using these probabilities and Equation 1.9, mutual information can be computed for each pair of dimensions of the dataset.

Some of the dimensions that exhibit high mutual information in the cars dataset are MPG and weight, weight and acceleration, horsepower and weight, etc. This is the expected behaviour, since, as we know, lighter cars have good fuel economy and better acceleration. Thus, mutual information provides an indication to the user of which axis pairs are likely to convey interesting information.

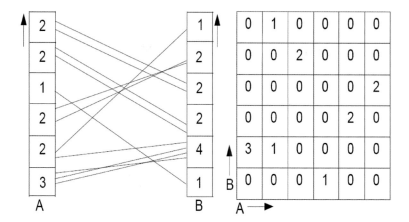

Figure 6.2 Example of the histogram computation on the parallel coordinates visualization [38].

Using this metric, the authors proposed to optimize the display for the analyst. In general, finding an optimal ordering of axes for parallel coordinates is NP-complete. Dasgupta and Kosara [38] suggest, considering the special properties of parallel coordinates, a branch-and-bound algorithm to find optimal solutions in much less time. Since most of the computations are performed on the histograms, the method is not very sensitive to the total number of data items, leading to a very efficient solution. First, a matrix with the MI value associated to all axis pairs is built. This computation is only performed once, and it is the only step that depends on the number of records in the dataset, since it requires the histogram computation. All subsequent steps are performed on the basis of this matrix and, thus, only depend on the number of dimensions. Then, the optimization is performed as a branch-and-bound algorithm that uses a priority queue and best-first search. A key issue in branch-and-bound implementations is how tightly the bounds can be estimated when the decision is made about whether to cull a sub-tree or not. In this case, these estimates are based on a full cost matrix, and are very precise. Finally, the axis ordering that maximizes the total sum of mutual information of axis pairs is obtained and shown to the data analyst.

Dasgupta and Kosara [38] also proposed another information-theoretic measure to quantify the quality of the visualization in parallel coordinates. In particular, they proposed to compute the entropy from a rendered parallel coordinates image, using the gray levels as the alphabet that is being transmitted. In this case, the probability of a gray level i is given by

$$p_i = \frac{n_i}{N}, \tag{6.4}$$

where n_i represents the number of pixels with gray level i and N the total number of pixels. From this probability, the entropy is given by

$$H = -\sum_{i=0}^{255} p_i \log p_i. \tag{6.5}$$

In parallel coordinates, a high entropy indicates a region with a large number of line crossings, but no inverse structure. On the contrary, an inverse relationship leads to a large amount of white pixels at the top and bottom of the region, which greatly reduces the entropy. While there is no simple connection between entropy and display structures, maximizing entropy generally leads to busy but very readable displays of the data. Typically, the polylines corresponding to each data sample are rendered with semitransparency, since this leads to better perceptual results due to less cluttering. Entropy can also be used in order to decide the α value (i.e., the degree of transparency) assigned to the polylines.

6.4   MAXIMUM ENTROPY SUMMARY TREES

One of the main problems related to data visualization is the excess of information. In this case, the representation of all the data is not possible due to the limitations of the canvas size, the cluttering, or the impossibility to interpret the data by the user. In this situation, a common strategy is to represent only the relevant information, while the irrelevant information is hidden. Information theory has also been used in the field of data visualization as a tool to determine the relevant data to be shown.

Karloff and Shirley [70] focused on the problem of how to simplify a very large node-weighted rooted tree by a summary tree in the most informative way. Summary trees are simplified trees that result from aggregating nodes of the original weighted tree, subject to certain constraints. In the summary tree definition, two types of contraction are allowed: First, subtrees are contracted to single nodes that represent the corresponding subtrees; and, second, multiple sibling subtrees (subtrees whose roots are siblings) are contracted to single nodes representing them. The node resulting from the latter is called the *other* group node. One constraint in the definition of summary trees is that each node has at most one child that is an *other* group node. Karloff and Shirley [70] suggest that the best choice among all the possible summary trees with a fixed number of nodes is the one that maximizes the entropy of the probability distribution associated with the summary tree. Formally, the definition of the entropy, $H(T)$, of a k-node tree T with node weights $W_1, W_2, \ldots, W_k$ is defined as:

$$H(T) = \sum_{i=1}^{k} p_i \log p_i, \tag{6.6}$$

where $p_i = W_i/W$ and W is the sum of all node weights. This definition

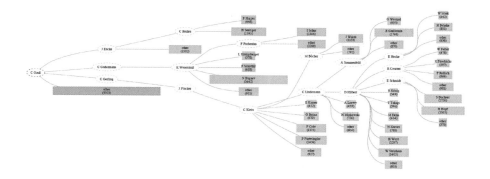

Figure 6.3 The maximum entropy 56-node summary tree of the math genealogy tree rooted at Carl Friedrich Gauss [70].

corresponds to the Shannon entropy (see Equation 1.1) where the probability density function is given by the normalized weight of each tree node. Figure 6.3 shows an example of a maximum entropy summary tree of a mathematical genealogy tree [40]. Some nodes are represented in their original form (ellipses), some others represent a subtree (rectangular shape with a name and a weight), and others represent the grouping of sibling subtrees (rectangular shape with the other label and the corresponding weight).

The construction of a summary tree S from the tree T can be defined in a more formal way [29]. Let T_v denote the subtree of T rooted at v. We name each node of S by the set of nodes of T that it represents and its weight is given by the sum of the weights of those nodes. The following comprises the possible summary trees for T_v: If T_v has just one node, the only summary tree is the singleton node $\{v\}$; otherwise, a summary tree for T_v is one of

1. a one-node tree $V(T_v)$ (the set of nodes in T_v); or

2. a singleton node $\{v\}$ and summary trees for the subtrees rooted at the children of v (and edges from $\{v\}$ to the roots of these summary trees); or

3. a singleton node $\{v\}$, a node $other_v$ representing a non-empty subset U_v of v's children and all the descendants of the nodes $x \in U_v$, and for each of v's children $x \notin U_v$ is a summary tree for T_x (and edges from $\{v\}$ to $other_v$ and to the roots of the summary trees for each T_x).

Figure 6.4(a) shows a 10-node tree example used to present some basic properties of the entropy measure on a summary tree. The weight of each node is shown in parenthesis and the total weight of the tree is 72.5. Thus,

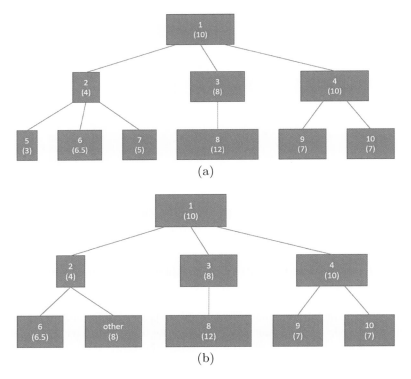

Figure 6.4 A 10-node tree example (with node weights in parenthesis) and a corresponding 9-node summary tree.

from Equation 6.6, its entropy is given by

$$
\begin{aligned}
H(T) &= -\frac{10}{72.5} \log \frac{10}{72.5} - \frac{4}{72.5} \log \frac{4}{72.5} - \cdots - \frac{7}{72.5} \log \frac{7}{72.5}. \\
&\cong 3.2188 \text{ bits.}
\end{aligned} \tag{6.7}
$$

Note that the maximum entropy value for a k-node tree is $\log_2 k$. This maximum value is achieved when all the nodes have the same weight. In this case, the maximum value is $\log_2 10 = 3.3219$, and the similarity between the real entropy and the maximum possible value indicates that this tree has nodes with similar weights (at least the node weights have the same order of magnitude). Thus, by maximizing the entropy of the summary tree, the obtained tree tends to have equally sized nodes, which seems a reasonable optimization strategy.

Another interesting property of the entropy measure can be seen from the graph S shown in Figure 6.4(b). A 9-node summary tree has been obtained from the graph in Figure 6.4(a), where the sibling nodes 5 and 7 have been aggregated into a single *other* node. If the entropy is computed in a manner similar to the previous example, $H(S) \cong 3.1135$ bits. It can be easily shown that the decrease of the tree entropy when two nodes v_i and v_j are aggregated into a single node v_{ij} is given by

$$
\begin{aligned}
\Delta H &= H(T) - H(S) \\
&= \frac{w_{ij}}{W} \log_2 \frac{w_{ij}}{W} - \frac{w_i}{W} \log_2 \frac{w_i}{W} - \frac{w_j}{W} \log_2 \frac{w_j}{W},
\end{aligned}
\tag{6.8}
$$

where W is the total weight of the tree nodes. Note the *locality* property of the loss of entropy, since it only depends on the weights of the nodes, but not on the other configuration of the tree.

6.4.1 How to Construct a Maximum Entropy Summary Tree

Cole and Karloff [29] proposed an algorithm to construct a maximum entropy summary tree based on dynamic programming. The first step consists of relabelling the nodes as $1, 2, \ldots, n$, with the root being node 1, the nodes at depth d getting consecutive labels, and the children of a node being labeled with increasing consecutive labels in nondecreasing size order. The *pseudo-entropy* $p\text{-}ent(S_v)$ of summary trees for S_v with nodes of weights $w_1, w_2, \ldots, w_k$ is defined as

$$
p\text{-}ent(S_v) = \sum_{i=1}^{k} \frac{w_i}{W} \log_2 \frac{w_i}{W},
\tag{6.9}
$$

where W is the total weight of T (and not of S_v). Note that this measure cannot be considered as an entropy measure since $\sum_{i=1}^{k} \frac{w_i}{W} \neq 1$. Clearly, if S_v is part of a summary tree S for T, then the nodes of S_v contribute $p\text{-}ent(S_v)$ to the entropy of S. Let $H(S_v)$ denote the entropy of the tree S_v. Then,

$$
\begin{aligned}
H(S_v) &= -\sum_i \frac{w_i}{W_v} \log \frac{w_i}{W_v} \\
&= -\left[\frac{W}{W_v} \sum_i \frac{w_i}{W} \log \frac{w_i}{W} + \sum_i w_i W \log \frac{W}{W_v} \right] \\
&= -\frac{W}{W_v} p\text{-}ent(S_v) - \log \frac{W}{W_v}.
\end{aligned}
\tag{6.10}
$$

Thus, the same tree optimizes the entropy and the pseudo-entropy. Cole and Karloff [29] proposed an algorithm based on dynamic programming. This algorithm starts from a set of small subtrees that minimizes the pseudo-entropy measure and adds new nodes in a way that the pseudo-entropy of the new subtree is still minimized. This procedure leads to the final optimal solution when all the nodes of the original tree have been considered.

6.5 MULTIVARIATE DATA EXPLORATION

In most real-world phenomena, the complex interactions between different variables are due to multiple factors. To gain an in-depth understanding of a scientific process, the relationship among the variables needs to be thoroughly investigated. However, the simultaneous exploration of several variables can be both tedious and confusing. For these reasons, exploration and discovery of relationships between different variables in multivariate datasets is one of the most challenging tasks in information visualization.

Biswas et al. [5] proposed an information-theoretic framework to guide the user at each step of the exploration process and to help him or her towards in-depth analysis of the datasets when multiple variables are involved. From an information-theoretic point of view, each variable in a multivariate data set can be seen as a random variable with its corresponding entropy, which measures the amount of information of the variable. Part of this information is shared by other variables, which can be characterized by the mutual information measure. From these basic ideas, the authors develop a general framework to characterize the variables and their relationships. The next sections describe the main parts of the framework.

6.5.1 Variable Entropy

The most basic measure of information is the Shannon entropy (see Equation 1.1). This is a measure of the uncertainty of a random variable and enables us to measure the information of each variable in the dataset.

In the multivariate datasets, the variables can be grouped on certain clusters based on a given criterion. In this case, entropy can also be used to compute the total amount of information in a cluster. For a collection of random variables, $X_1, \ldots, X_n$, the total information content among the variables can be expressed by the joint entropy (see Equation 1.3) which is defined as

$$H(X_1, \ldots, X_n) = -\sum_{x_1 \in \mathbb{X}_1} \cdots \sum_{x_n \in \mathbb{X}_N} p(x_1, \ldots, x_n) \log p(x_1, \ldots, x_n).$$

$$(6.11)$$

Thus, the joint entropy can be applied to measure the total uncertainty within each group, where the joint probability distribution of the variables is used. This allows us to compute the relative importance of the groups based on their uncertainty. The groups can be selected depending on their uncertainty and the individual variables inside the selected groups that need to be analyzed.

In information theory, given a group of variables, the conditional entropy (see Equation 1.4) is used to quantify the information gain about the system when some of its variables are known. From a set of n variables $\{X_1, \ldots, X_n\}$, if another set of m variables $\{X_{k1}, \ldots, X_{km}\}$ is known, the amount of uncertainty remaining on the system is given by

$$H(X_1, \ldots, X_n | X_{k1}, \ldots, X_{km}) = H(X_1, \ldots, X_n, X_{k1}, \ldots, X_{km})$$
$$- H(X_{k1}, \ldots, X_{km}).$$
$$(6.12)$$

This provides a useful measure to identify variables inside subgroups that

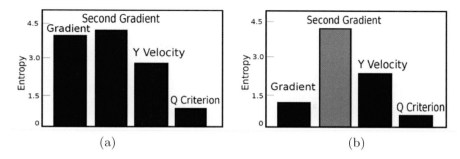

Figure 6.5 Change in uncertainty of the variables due to the variable selection. The left plot shows the uncertainty remaining in the variables before selection and the right plot shows the uncertainty remaining in the variables after selection of the Second Gradient [5].

have a larger contribution to the total uncertainty of a group. In addition, this metric allows us to select those variables that represent the uncertainty of the whole subgroup by their information content.

A simple example of how the selection of one variable can reduce the uncertainty about other variables can be seen in Figure 6.5 (from [5]). This example uses four variables of the Plume dataset, which is obtained from a simulation of the thermal downflow plumes on the surface of the sun with $126 \times 126 \times 512$ grid points. Figure 6.5(a) shows the individual entropies of each variable of the dataset. If we want to first explore the highest uncertain variables, then the order of selection would be Second Gradient, Gradient, Y Velocity, and Q Criterion. But once the Second Gradient field has been selected, it will now have an impact on our knowledge about the other variables and the conditional entropy has to be used instead of the individual entropy. Figure 6.5(b) shows the conditional entropy of each variable conditioned to the Second Gradient (in the case of the Second Gradient, the individual entropy is shown). Hence, the second variable to be chosen should be Y Velocity, as the uncertainty about the Gradient field, which was the second highest individual entropy value, has been reduced due to its higher correlation with the Second Gradient field.

6.5.2 Mutual Information between Variables

In multivariate analysis, the correlation between different variables is a well-researched topic and several metrics have been proposed. In information theory, mutual information between two random variables is the measure of the information overlap or the correlation between the variables. For two random variables X and Y, mutual information $I(X; Y)$ (see Equation 1.9) is defined

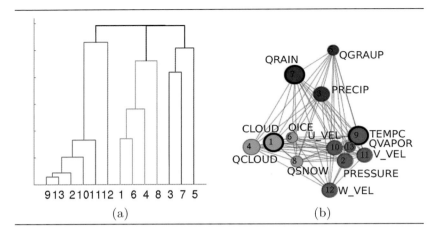

(a)   (b)

Figure 6.6 The MI relationship between variables shown as a dendrogram (a) and as a graph layout (b) [5].

as

$$I(X;Y) \quad = \quad \sum_{x \in \mathbb{X}} \sum_{y \in \mathbb{Y}} p(x,y) \log \frac{p(x,y)}{p(x)p(y)}. \tag{6.13}$$

As a correlation metric, mutual information has an advantage over the correlation coefficient metric, as it can measure non-linear relationships as well. Mutual information between two variables quantifies the informativeness of one variable about the other variable.

Biswas et al. [5] proposed two different ways to visualize the relationship between the variables in a dataset. First, for each pair of variables, the distance between the variables is defined as the inverse of the mutual information between them. Considering all the variables of the system, a graph $G(V, E)$ is constructed to delineate the system of variables. Each node $v \in V$ represents a variable, and each undirected edge $e \in E$ represents the mutual information between the two variables. A hierarchical clustering is applied on this graph to decompose it into different groups. In a bottom-up clustering approach, each node represents a leaf of the cluster tree and each of them starts out as a cluster. Then, the new groups are formed via a greedy algorithm that merges the two clusters with most similarity to move up the cluster tree one level.

To show this approach, the Hurricane Isabel data set was used. The Hurricane Isabel data was produced by the Weather Research and Forecast (WRF) model, courtesy of NCAR and the U.S. National Science Foundation (NSF). This model consists of thirteen variables. Figure 6.6(a) shows the dendrogram of the clustered graph, which reveals the hierarchy of the clusters. It is evident from the figure that there exist three major subgroups that form clusters according to the information overlap. In Figure 6.6(b), an alternative graph view of the system is provided. The layout of the graph is generated by a force-directed algorithm, based on a dynamic spring system where the total energy of the system is minimized. In this model, the attractive force between two nodes i and j (that represent data variables) is given by

$$F_{ij} \propto \frac{1}{d_{ij}^2},$$

(6.14)

where d_{ij} is the mutual information between the variables i and j. As it can be seen in Figure 6.6, the force-directed layout and the hierarchical clustering dendrogram show consistent results.

6.5.3 Specific Information

Another interesting analysis on multivariate datasets is how a given value on one variable is related to another variable. To quantify this, two measures based on different decompositions of mutual information were proposed [5]. The information measure associated with a specific scalar value x, about another random variable Y is called *specific information* (see Section 1.3). In this case, the variable X is called the *reference variable*. From the several ways to decompose mutual information, the two proposed by DeWeese and Meister [41] are the most natural decompositions. These two specific information metrics were named *surprise* and *predictability* by Bramon et al. [10]. Next, we describe their main features:

- **Surprise** I_1. The definition of *surprise* I_1 (see Equation 1.18) is given by

$$I_1(x; Y) = \sum_{y \in \mathbb{Y}} p(y|x) \log \frac{p(y|x)}{p(y)}.$$

(6.15)

This measure expresses the surprise about Y from observing the value x. The surprise always takes positive values and can be seen as the Kullback–Leibler distance between $p(Y|x)$ and $p(Y)$.

- **Predictability** I_2. The definition of *predictability* I_2 (see Equation 1.19) is given by

$$
\begin{aligned}
I_2(x; Y) &= H(Y) - H(Y|x) \\
&= -\sum_{y \in \mathbb{Y}} p(y) \log p(y) + \sum_{y \in \mathbb{Y}} p(y|x) \log p(y|x).
\end{aligned}
$$

(6.16)

This measure expresses the change in uncertainty about Y when x is observed. Note that $I_2(x; Y)$ can take negative values, since certain observations of X can increase our uncertainty about the variable Y.

These specific information measures provide us with the tools for classifying the individual scalars of a variable. Given two variables of a dataset, each of them can be considered as a random variable and, for each scalar value of the reference variable chosen between the two variables, the I_1 and I_2 metrics can be computed. To guide the exploration of the dataset, the scalars that have a high I_1 value are identified as the surprising ones and are further classified by their I_2 values.

Since the specific information metric I_2 represents the uncertainty or predictability factor and I_1 describes the "surprise" of the scalar value, to show the scalar values that correspond to higher uncertainty in the other variable, a derived metric I_{uncert} is formulated such that

$$I_{uncert} = \frac{I_1}{I_2}. \qquad (6.17)$$

High values of I_{uncert} are obtained for high I_1 values and low I_2 values. Similarly, for the scalar values that correspond to lower uncertainty in the other variable, a derived metric I_{cert} is given by

$$I_{cert} = I_1 * I_2. \qquad (6.18)$$

In this case, high values of I_{cert} are obtained for high values of both I_1 and I_2 measures.

Biswas et al. [5] suggest using the I_{cert} and I_{uncert} metrics to choose the scalars from a parallel coordinate visualization. Figure 6.7 presents two exploration tools for the Hurricane Isabel dataset with the parallel coordinates technique based on these measures. Figure 6.7(a) shows the selection of scalar values using I_{cert} where a small range of scalars of pressure maps to a small range of scalars in temperature. Conversely, Figure 6.7(b) shows the selection of scalars using I_{uncert} where a small range of scalar values of pressure maps to a much larger range of scalar values on the temperature axis. Similarly, other variables of the system can be used for exploration with effective results.

6.6 TIME-VARYING MULTIVARIATE DATA

When dealing with time-varying multivariate data, the detection of relationships between variables is a very important issue, similar to the problem that has been described in the previous section. But in this case, another important factor is to determine causal relationships among multiple variables, considering now the arrow of time. Commonly used techniques for the estimation of dependencies are linear cross-correlation and mutual information. However, these measures share the property of being symmetric and therefore are not suited for assessing causality within relationships. Wang et al. [138] proposed a new visualization framework to study the directional aspect of interactions. This framework is based on the transfer entropy proposed by Schreiber [103] to quantify the information flow between time series. With minimal assumptions

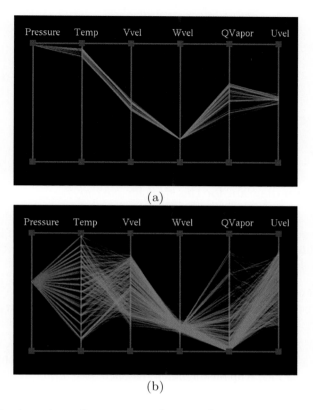

Figure 6.7 Exploration of pressure values and temperature variables inside the subgroup of Isabel data [5].

about the dynamics of the systems and the nature of their coupling (for instance, no linear relationship is assumed), this information-theoretic measure can quantify the exchange of information between two systems, separately for each direction.

6.6.1 Transfer Entropy

In this framework, every time series can be seen as a stationary Markov process of order k, that is, the probability of every state x_{n+1} depends only on the previous k states. Formally, a stationary process $\{X_1, X_2, \ldots, X_n, X_{n+1}, \ldots\}$ is a Markov process of order k if

$$p(x_{n+1}|x_n, \ldots, x_1) = p(x_{n+1}|x_n, \ldots, x_{n-k+1}). \tag{6.19}$$

Section 1.4 describes, in more detail, other features of Markov processes. Let us denote $x_n^{(k)} = (x_n, \ldots, x_{n-k+1})$ for words of length k, where the subscript

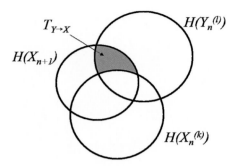

Figure 6.8 Venn diagram for the transfer entropy.

denotes the state (or time step) and the superscript denotes the length of the states (or time steps) considered. The transfer entropy between two variables X and Y is defined by Schreiber [103] as

$$
\begin{aligned}
T_{Y \to X} &= H(X_{n+1}|X_n^{(k)}) - H(X_{n+1}|X_n^{(k)}, Y_n^{(l)}) \\
&= \sum p(x_{n+1}, x_n^{(k)}, y_n^{(l)}) \log \frac{p(x_{n+1}|x_n^{(k)}, y_n^{(l)})}{p(x_{n+1}|x_n^{(k)})},
\end{aligned} \tag{6.20}
$$

where $T_{Y \to X}$ denotes the influence of Y on X. Note that this definition can be seen as the Kullback–Leibler distance (see Equation 1.7) between $p(X_{n+1}|X_n^{(k)}, Y_n^{(l)})$ and $p(X_{n+1}|X_n^{(k)})$, that is, the difference between the probabilities of X at state $n + 1$ when all the previous k and l states of X and Y are known, respectively, and the probabilities when only the previous k states of X are known. In other words, it can be seen as the reduction of uncertainty on X when the past of both X and Y are known with respect to the uncertainty on X when only the past of X is known. This measure can be represented by the Venn diagram shown in Figure 6.8. Note also that $T_{Y \to X}$ is explicitly non-symmetric under the exchange of X and Y (a similar expression exists for $T_{X \to Y}$) and, thus, it can be used to detect the directed exchange of information between the two time series.

The most natural choices for l are $l = k$ (the same number of time steps is considered for both X and Y) or $l = 1$ (only one time step for Y is considered at a time). Usually, the latter is preferable due to its lower computational cost, but long time dependencies are not taken into account. When $k = 1$ and $l = 1$, the transfer entropy can be written as

$$
\begin{aligned}
T_{Y \to X} &= H(X_{n+1}|X_n) - H(X_{n+1}|X_n, Y_n) \\
&= H(X_{n+1}, X_n) + H(X_n, Y_n) - H(X_n) - H(X_{n+1}, X_n, Y_n).
\end{aligned} \tag{6.21}
$$

6.6.2   Visualization of Information Transfer

Wang et al. [138] present a new way to visualize information transfer. The left plot of Figure 6.9 shows an example of a pair-wise representation of this visualization. For each variable V_i, a circle denotes the total outgoing and incoming influence at a certain time step. The size of the circle shows the total amount of influence with green/orange for outgoing/incoming influence. The edge e_{ij} indicates the influence between variables V_i and V_j. The transfer entropy $T_{V_i \to V_j}$ is mapped to the edge width at vertex v_i and $T_{V_i \to V_j}$ to v_j. Green/orange indicates more outgoing/incoming influence. The color saturation is adjusted according to the absolute difference between the two transfer entropies so that pairs of variables with larger difference would stand out. Edge width and color in between are linearly interpolated. To reduce the occlusion among edges, all edges are sorted in the decreasing order of their average thickness and then the edges are drawn accordingly. On the right plot of Figure 6.9, a circular graph is plotted to show information transfer for the Hurricane Isabel dataset at time step 4 (from [138]). This graph shows the relationship between nine variables: QC (cloud moisture mixing ratio), QI (cloud ice mixing ratio), QG (Graupel mixing ratio), QR (rain mixing ratio), QS (snow mixing ratio), QV (water vapor mixing ratio), PR (pressure), TC (temperature), and WS (wind speed magnitude). Such circular graphs are intuitive for inferring the relationship among all pairs of variables, as a group and as individuals. Since every time step corresponds to such a graph, we can produce a time-varying graph showing the information transfer. These graphs can answer questions such as which variable has the most total outgoing and incoming influence (find the largest circle), which variable's total outgoing influence is more than its total incoming influence and by how much (see the green and orange portions of the circles), and, for a pair of variables, which variable influences the other more and by how much (compare edge color and width difference at both ends). For instance, it can be seen that the wind speed magnitude is a variable with high influence, where the incoming one is much greater than the outgoing one, which indicates that it is more a consequence of the other variables than the cause. On the contrary, the pressure and temperature variables can be considered much more a cause than a consequence, since they both have much more outgoing influence than incoming influence. From the plot, it can also be seen that these three variables are highly related, since the edges between them are wide.

6.7   PRIVACY AND UNCERTAINTY

In some applications where sensitive data are visualized, there is a balance between conveying information and maintaining privacy. It is necessary for the visualization designer to beware of the possibility that an attacker may be able to gain sensitive information from visualization. For example, from a parallel coordinate plot of anonymized personal records, an attacker may

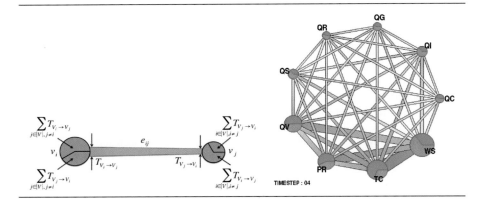

Figure 6.9 Visualization of the transfer entropy. Left: pair-wise example. Right: circular graph to show information transfer for the hurricane dataset [138].

start from a position on an axis, through which only a few record lines pass. By following these few lines, the attackers may find all individual attribute values of the corresponding records. With some additional effort, such as deductive elimination or exhaustive search, the attacker may be able to identify supposedly anonymized individuals. *Privacy-preserving visualization* is an approach of visual design that introduces uncertainty deliberately to mask the actual values in the original data records. For example, using common visual channels [17], actual data values can be masked by mapping multiple values to the same color (e.g., in pixel-based visualization), the same shape (e.g., in glyph-based visualization), or the same position (e.g., in parallel coordinate plots).

In general, increasing the amount of visual uncertainty in a visualization will increase the privacy of the visualization, but decrease its utility. In other words, privacy and utility are functions of visual uncertainty. Information theory can thus be used to measure the amount of uncertainty in a visualization [37], since this process can be viewed as a communication channel from the data space to the perceptual and cognitive space (or mental space) of the user [63, 93].

6.7.1 Metrics for Uncertainty

In Section 6.3, we described some screen-space metrics for measuring visual uncertainty, such as pattern complexity in parallel coordinates [38]. Such metrics were adapted by Dasgupta et al. for measuring uncertainty in privacy-preserving visualizations [37].

Encoding uncertainty serves as the initial defensive mechanism against at-

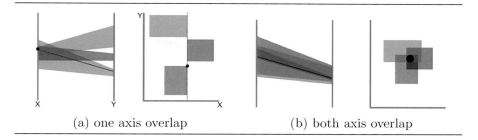

(a) one axis overlap   (b) both axis overlap

Figure 6.10 Examples of how overlaps lead to uncertainty. (a) Overlap on one axis and (b) on both axes [37].

tackers with no background knowledge about the data. *Binning* can result in reduction of precision, while clustering can lead to loss of granularity. These encoding strategies make it difficult for an attacker to guess the exact value and number of data points within a bin or a cluster. However, when an attacker has some partial background knowledge about the data, the different components in a visualization will have different capabilities of confusing the attacker. For example, when an attacker knows about the existence of a particular data point among data records being visualized, the attacker may start the process of invading privacy by associating the data point with a cluster. In such a case, overlapping clusters make the association more difficult. As shown in Figure 6.7.1, overlapping on the X axis can introduce uncertainty undesirable to an attacker knowing a data point on X (marked by the black dot). Overlapping on both the X and Y axes bring about further difficulties for an attacker knowing two data points in a record.

Dasgupta et al. [37] proposed several metrics of privacy in cluster-based parallel coordinate plots. Among these metrics, there is an information-theoretic measure.

Consider an axis that has h pixel bins, in a privacy-preserving parallel coordinates visualization. Consider n_c clusters that intersect with this axis. Each cluster may span over one or more bins, while each bin may intersect with zero, one, or several clusters. As identifying an empty bin is trivial, the uncertainty is thus associated with those bins that intersect with one or more clusters. Assume that the attacker has no a priori knowledge about any cluster, so the probability of making a correct guess of the association between a bin and a cluster is independent and identically distributed.

Let a_i be the number of clusters intersecting with bin b_i, where $i \in [0, h-1]$. Given a cluster, C_t, the probability for identifying this cluster at bin b_i is thus

$$p_{t,i} = \begin{cases} 0 & \text{if } C_t \text{ does not intersect with } b_i \\ 1/a_i & \text{if } C_t \text{ intersects with } b_i. \end{cases}$$

The *self-information* of C_t at bin b_i is defined as

$$s_{t,i} = \begin{cases} 0 & \text{if } p_{t,i} = 0 \\ -\log p_{t,i} & \text{if } p_{t,i} > 0. \end{cases}$$

$s_{t,i}$ is also referred to as the *surprisal value*. The larger the value is, the more surprised we would be that it is discovered, and hence the more private it is.

We can consider collectively the surprisal value of C_t across the axis, i.e., at all bins with which C_t intersects. Dasgupta et al. [37] proposed to sum up all non-zero $s_{t,i}$ weighted by $p_{t,i}$. Since an attacker likely considers that a bin with fewer clusters is an easier target, the weight $p_{t,i}$ represents the likelihood that bin b_i being attacked in the context of C_t. This results in a privacy indicator ϕ_t for C_t:

$$\phi_t = \sum_{i=0}^{h-1} p_{t,i} s_{t,i} = -\sum_{i=0}^{h-1} p_{t,i} \log p_{t,i}, \tag{6.22}$$

which appears similar to Shannon entropy, but it does not have the same meaning because $p_{t,i}$ is not defined as a probability mass function for a set. Nevertheless, the higher ϕ_t is, the higher the privacy is about C_t at this axis. A better definition of weight is to normalize $p_{t,i}$, such that

$$p'_{t,i} = \frac{p_{t,i}}{\sum_{i=0}^{h-1} p_{t,i}}. \tag{6.23}$$

Alternatively, one may combine all non-zero $s_{t,i}$ using their minimum, average, or summation.

The collective uncertainty of the axis with all clusters can therefore be summed as

$$\Phi_{total} = \sum_{t=1}^{n_c} \phi_t. \tag{6.24}$$

Because this sum depends on the number of clusters n_c, one may find it more intuitive to use the following relative form:

$$\Phi_{rel} = \frac{\sum_{t=1}^{n_c} \phi_t}{n_c \phi_{max}}, \tag{6.25}$$

where ϕ_{max} is the maximum value for ϕ_t, which is associated with a situation where every cluster spans over every pixel bin such that $p_{t,i} = 1/n_c$ for every cluster and every bin.

6.8 MUTUAL INFORMATION DIAGRAM

In the previous sections of this chapter, we have shown how information theory can be used to model and assess the information visualization pipeline. In this section, we present how information-theoretic measures can be used to design a new visualization technique.

As we previously mentioned, in multivariate data analysis, a key point is to determine the relationship between variables. In particular, in uncertainty studies, a major task is the comparison of different models to determine whether they are effective in explaining the observed data. This is mainly done via statistical analysis by quantifying the correlation between the model predicted values and the observed ones. From the visualization perspective, the most used technique to deal with this problem is via scatter plot matrices. But this technique becomes practically unusable when the number of variables is relatively large.

In order to overcome this problem, Taylor [120] proposed a new type of diagram, usually called the Taylor diagram, that maps several statistics of the relationship between two variables in a single point on this diagram, allowing an easy way to compare multiple models or variants of a model. This plot describes the variable relationship in terms of their variances, their correlation, and their centered root mean square difference. This plot is widely used in geophysics and climate change research fields. However, not all relationships can be explained in terms of variance and correlation alone. In many cases, variables exhibit non-linear dependence, a fact that cannot be correctly identified using linear correlation. In other cases, in the presence of outliers, two variables may exhibit low correlation, while otherwise being relatively similar. Correa and Lindstrom [31] proposed a new diagram, called the *mutual information diagram*, based on information-theoretic measures to quantify the relationship between the variables, taking into account non-linear relationships and obtaining more robust results in the presence of outliers. The next sections describe the main features of the Taylor diagram and the mutual information diagram.

6.8.1 The Taylor Diagram

The Taylor diagram provides a graphical representation that allows us to visually compare the statistical relationship between random variables. An example of a Taylor diagram is shown in Figure 6.11(a). This plot is based on three statistical measures between the two variables: the standard deviation of each variable, their correlation, and their centered root mean square difference. Consider two discrete random variables X and Y, with means μ_X and μ_Y and standard deviations σ_X and σ_Y, respectively. Let R_{XY} denote Pearson's correlation coefficient

$$R_{XY} = \frac{cov(X, Y)}{\sigma_X \sigma_Y}, \tag{6.26}$$

where

$$cov(X, Y) = \frac{1}{n} \sum_{i=1}^{n} (x_i - \mu_X)(y_i - \mu_y) \tag{6.27}$$

is the covariance between X and Y and n is the number of samples.

The centered root mean squared (RMS) difference between X and Y is given by

$$RMS(X,Y) \; = \; \sqrt{\frac{1}{n} \sum_{i=1}^{n} ((x_i - \mu_X)(y_i - \mu_y))^2}. \qquad (6.28)$$

These three measures fulfill the following property

$$\begin{aligned} RMS(X,Y)^2 \; &= \; \sigma_X^2 + \sigma_Y^2 - 2\sigma_X \sigma_Y R_{XY} \\ &= \; \sigma_X^2 + \sigma_Y^2 - 2cov(X,Y) \\ &= \; cov(X,X) + cov(Y,Y) - 2cov(X,Y). \qquad (6.29) \end{aligned}$$

From this property, a representation of these measures can be achieved by using a polar coordinate system and the law of cosinus (i.e., $c^2 = a^2 + b^2 - 2ab\cos\theta$). As can be seen in Figure 6.11(a), the relationship between X and Y is shown by a point drawn with polar coordinates of $r = \sigma_Y$ and $\theta = cos^{-1}(R_{XY})$. A point on the X axis at position σ_X represents the observed data variability. The Euclidian distance between these two points is given by $RMS(X,Y)$. Remember that this plot is usually used to show the fitness between the observed data and the predicted one by a given model and, thus, X represents the model's observed data and Y the model's predicted data. As the performance of each model is represented by a single point in the Taylor diagram, this plot is typically used to compare multiple models by adding a point for each model.

The half Taylor plot (like the one in Figure 6.11(a)), which is represented in the first quadrant, can only show positive correlations (i.e., $R_{XY} \in [0,1]$). Since R_{XY} can take values in the range $[-1,1]$, the full Taylor diagram, which uses two quadrants, is able to represent both positive and negative correlations.

6.8.2   Mutual Information Diagram

As we previously mentioned, one of the limitations of the Taylor diagram is that the relationship between the variables is measured in terms of linear correlation. However, this may not be the case for many distributions. Correa and Lindstrom [31] proposed the *mutual information diagram*, which extends the Taylor diagram by using information-theoretic measures.

To define this diagram, the authors used the following relationships between basic information-theoretic measures. Variation of information (VI) [85] between two distributions X and Y is defined as

$$\begin{aligned} VI(X,Y) \; &= \; H(X|Y) + H(Y|X) \\ &= \; H(X) + H(Y) - 2I(X;Y) \\ &= \; I(X;X) + I(Y;Y) - 2I(X;Y). \qquad (6.30) \end{aligned}$$

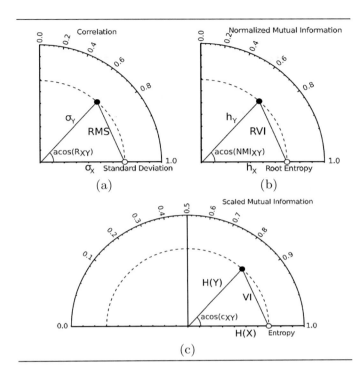

Figure 6.11 The Taylor diagram (a) and two versions of the mutual information diagram: (b) the RVI-based diagram and (c) the VI-based diagram [31].

This measure is a metric, and satisfies properties such as non-negativity, symmetry, and triangle inequality [85]. Correa and Lindstrom [31] show that $RVI(X,Y) = \sqrt{VI(X,Y)}$ is also a metric. Another measure defined for convenience is the normalized mutual information (NMI), given by

$$NMI(X,Y) \;=\; \frac{I(X;Y)}{\sqrt{H(X)H(Y)}}. \tag{6.31}$$

This measure takes values in $[0,1]$, taking the value 0 if and only if X and Y are independent. This normalized version of mutual information has been previously used in other works [116]. From these measures, the following equation can be easily obtained

$$
\begin{aligned}
RVI(X,Y)^2 &= H(X) + H(Y) - 2I(X;Y) \\
&= RH(X)^2 + RH(Y)^2 - 2RH(X)RH(Y)\frac{I(X;Y)}{RH(X)RH(Y)} \\
&= RH(X)^2 + RH(Y)^2 - 2RH(X)RH(Y)NMI(X,Y),
\end{aligned}
\tag{6.32}
$$

where $RH(X) = \sqrt{H(X)}$ and $RH(Y) = \sqrt{H(Y)}$.

By comparing Equations 6.29 and 6.32, the following mapping can be done in order to use information-theoretic measures in the Taylor diagram:

- Error $RMS(X,Y)$ ⇔ root variation of information $RVI(X,Y)$

- Variance σ_X^2 ⇔ entropy $H(X)$

- Covariance $cov(X,Y)$ ⇔ mutual information $I(X;Y)$

- Correlation R_{XY} ⇔ normalized mutual information $NMI(X,Y)$

Thus, from this mapping, a new mutual information diagram, the RVI-based diagram [31], can be defined as shown in Figure 6.11(b). In this case, the position in polar coordinates is given by the square root of the entropy $RH(Y)$ at an angle defined by $\theta = cos^{-1}(NMI(X,Y))$. Note that this diagram can only be represented by one quadrant since $NMI(X;Y)$ takes values in $[0, 1]$.

Note that this latter diagram is based on the square root of the entropy, which is not a natural measure. Another version of the mutual information diagram, called the VI-based diagram, was proposed by Correa and Lindstrom [31]. The new diagram is based on the following equation:

$$
\begin{aligned}
VI(X,Y)^2 &= (H(X) + H(Y) - 2I(X;Y))^2 \\
&= H(X)^2 + H(Y)^2 - 2H(X)H(Y)c_{XY},
\end{aligned}
\tag{6.33}
$$

where c_{XY} is defined as

$$
c_{XY} = 2I(X;Y)\frac{H(X,Y)}{H(X)H(Y)} - 1.
\tag{6.34}
$$

Note that c_{XY} is a biased and scaled version of the mutual information and takes values in $[-1, 1]$, taking the value -1 if and only if X and Y are independent. In fact, the term $I(X;Y)\frac{H(X,Y)}{H(X)H(Y)}$, called the *scaled mutual information* $SMI(X,Y)$ [31], can be seen as a normalization of the mutual information and takes values between 0 and 1. In this case the mapping is as follows:

- Error $RMS(X,Y)$ ⇔ variation of information $VI(X,Y)$

- Standard deviation σ_X ⇔ entropy $H(X)$

- Correlation R_{XY} ⇔ c_{XY}

The corresponding plot is shown in Figure 6.11(c). In this case, two quadrants are needed since the measure $c_{XY} \in [-1, 1]$.

6.8.3 Properties of the MI Diagram

As shown in [31], the mutual information (MI) diagram has several advantages with respect to the standard Taylor diagram:

1. Mutual information-based measures are able to capture both linear and non-linear correlations.

2. Entropy and mutual information are less sensitive to outliers than the standard deviation and Pearson's correlation coefficient.

3. Information-based measures can handle with both numerical and categorical data while the Taylor diagram can only be constructed with numerical data.

On the other hand, there are some limitations on the MI diagram. First, the estimation of entropy and mutual information can be done using different techniques, such as histogram binning or using kernel estimates. These methods have certain parameters that influence the final measure value. The authors analyze the impact of these parameters and state that the relative comparisons between the methods are kept but not the values themselves. An aspect that was not analyzed in [31] is the fact that kernel estimates give an estimation of the continuous version of the entropy and mutual information measures. While continuous mutual information is defined as a positive measure, this is not the case for continuous entropy. Note that the standard MI diagram is not able to represent negative values of entropy. Another limitation of the diagram is that it does not distinguish between positive and negative correlations, which is an important point in certain applications.

6.9 SUMMARY

Information visualization is the set of techniques that provides visual representations of abstract data designed to strengthen human cognition. In this chapter, we presented several approaches that use information-theoretic measures to assess the information visualization process. First, we described some approximations of the theoretical foundations of the information visualization process based on information theory. We showed information-theoretic measures to quantify the quality of the visualization, giving special attention to the parallel coordinates technique. We explained the construction of summary trees based on the entropy measure and showed the use of information-theoretic measures to analyze the relationship between variables in multivariate datasets, focusing on the case of time-varying data. We also focused on measures to quantify the visualization uncertainty for privacy preservation. Finally, we described the mutual information diagrams.

FURTHER READING

Biswas, A., Dutta, S., Shen, H.-W., and Woodring, J. (2013). An information-aware framework for exploring multivariate data sets. *IEEE Transactions on Visualization and Computer Graphics*, 19(12):2683–2692.

Chen, M. and Jänicke, H. (2010). An information-theoretic framework for visualization. *IEEE Transactions on Visualization and Computer Graphics*, 16(6):1206–1215.

Cole, R. and Karloff, H. (2014). Fast algorithms for constructing maximum entropy summary trees. In Javier Esparza, Pierre Fraigniaud, Thore Husfeldt, and Elias Koutsoupias, editors, *Automata, Languages, and Programming*, volume 8572 of *Lecture Notes in Computer Science*, 332–343, Springer, Berlin, Heidelberg.

Correa, C. and Lindstrom, P. (2013). The mutual information diagram for uncertainty visualization. *International Journal for Uncertainty Quantification*, 3(3):187–201.

Dasgupta, A., Chen, M., and Kosara, R. (2013). Measuring privacy and utility in privacy-preserving visualization. *Computer Graphics Forum*, 32(8):35–47.

Dasgupta, A. and Kosara, R. (2010). Pargnostics: Screen-space metrics for parallel coordinates. *IEEE Transactions on Visualization and Computer Graphics*, 16(6):1017–1026.

Karloff, H. and Shirley, K.E. (2013). Maximum entropy summary trees. *Computer Graphics Forum (Proc. EuroVis)*, 32(3):71–80.

Purchase, H.C., Andrienko, N., Jankun-Kelly, T.J., and Ward, M. (2008). Theoretical foundations of information visualization. In Andreas Kerren, John T. Stasko, Jean-Daniel Fekete, and Chris North, editors, *Information Visualization*, volume 4950 of *Lecture Notes in Computer Science*, 46–64, Springer, Berlin, Heidelberg.

Wang, C., Yu, H., Grout, R.W., Ma, K.-L., and Chen, J.H. (2011). Analyzing information transfer in time-varying multivariate data. In *Proceedings of IEEE Pacific Visualization Symposium 2011*, 99–106.

Bibliography

[1] Carlos Andújar, Pere-Pau Vázquez, and Marta Fairén. Way-finder: Guided tours through complex walkthrough models. *Computer Graphics Forum*, 23(3):499–508, September 2004.

[2] Robert B. Ash. *Information Theory*. Dover Publications, Inc., New York, 1965.

[3] Pierre Barral, Guillaume Dorme, and Dimitri Plemenos. Visual understanding of a scene by automatic movement of a camera. In *Proceedings of the International Conference GraphiCon'99*, Moscow, Russia, August–September 1999.

[4] A. Bhattacharyya. On a measure of divergence between two statistical populations defined by their probability distributions. *Bulletin of the Calcutta Mathematical Society*, 35:99–109, 1943.

[5] Ayan Biswas, Soumya Dutta, Han-Wei Shen, and Jonathan Woodring. An information-aware framework for exploring multivariate data sets. *IEEE Transactions on Visualization and Computer Graphics*, 19(12):2683–2692, 2013.

[6] U. D. Black. *Data Networks: Concepts, Theory and Practice*. Prentice-Hall, Upper Saddle River, NJ, 1989.

[7] Xavier Bonaventura, Jianwei Guo, Weiliang Meng, Miquel Feixas, Xiaopeng Zhang, and Mateu Sbert. 3D shape retrieval using viewpoint information-theoretic measures. *Computer Animation and Virtual Worlds*, 26(2):147–156, 2015.

[8] Udeepta D. Bordoloi and Han-Wei Shen. View selection for volume rendering. In *Proceedings of IEEE Visualization*, pages 487–494, 2005.

[9] R. Borgo, J. Kehrer, D. H.S. Chung, E. Maguire, R. S. Laramee, H. Hauser, M. Ward, and M. Chen. Glyph-based visualization: Foundations, design guidelines, techniques and applications. In *Eurographics State of the Art Reports*, pages 39–63. Eurographics Association, May 2013.

[10] Roger Bramon, Imma Boada, Anton Bardera, Joaquim Rodriguez, Miquel Feixas, Josep Puig, and Mateu Sbert. Multimodal data fusion based on mutual information. *IEEE Transactions on Visualization and Computer Graphics*, 18(9):1574–1587, 2012.

[11] Roger Bramon, Marc Ruiz, Anton Bardera, Imma Boada, Miquel Feixas, and Mateu Sbert. An information-theoretic observation channel for volume visualization. *Computer Graphics Forum*, 32(3pt4):411–420, 2013.

[12] Roger Bramon, Marc Ruiz, Anton Bardera, Imma Boada, Miquel Feixas, and Mateu Sbert. Information theory-based automatic multimodal transfer function design. *IEEE Journal of Biomedical and Health Informatics*, 17(4):870–880, 2013.

[13] Stefan Bruckner and Torsten M"oller. Isosurface similarity maps. *Computer Graphics Forum*, 29(3):773–782, 2010. EuroVis 2010 Best Paper Award.

[14] Jacob Burbea and C. Radhakrishna Rao. On the convexity of some divergence measures based on entropy functions. *IEEE Transactions on Information Theory*, 28(3):489–495, May 1982.

[15] Daniel A Butts. How much information is associated with a particular stimulus? *Network: Computation in Neural Systems*, 14:177–187, 2003.

[16] Wenli Cai and Georgios Sakas. Data intermixing and multi-volume rendering. *Computer Graphics Forum*, 18(3):359–368, 1999.

[17] S.K. Card and J. Mackinlay. The structure of the information visualization design space. In *Proceedings of the IEEE Symposium on Information Visualization, 1997.*, pages 92–99, Oct 1997.

[18] Pascual Castelló, Mateu Sbert, Miguel Chover, and Miquel Feixas. Viewpoint entropy-driven simplification. In *Proceedings of the 15th International Conference in Central Europe on Computer Graphics, Visualization and Computer Vision (WSCG 2007)*, pages 249–256, University of West Bohemia, Plzen, Czech Republic, 2007. Václav Skala, UNION Agency.

[19] Pascual Castelló, Mateu Sbert, Miguel Chover, and Miquel Feixas. Viewpoint-driven simplification using mutual information. *Computers & Graphics*, 32(4):451–463, 2008.

[20] Ming-Yuen Chan, Huamin Qu, Yingcai Wu, and Hong Zhou. Viewpoint selection for angiographic volume. In *Advances in Visual Computing*, volume 4291 of *Lecture Notes in Computer Science*, pages 528–537. Springer, Berlin, Heidelberg, 2006.

[21] A. H. Charles and T. A. Porsching. *Numerical Analysis of Partial Differential Equations*. Prentice Hall, Englewood, NJ, 1990.

[22] M. Chen, D. Ebert, H. Hagen, R. S. Laramee, R. van Liere, K.-L. Ma, W. Ribarsky, G. Scheuermann, and D. Silver. Data, information and knowledge in visualization. *IEEE Computer Graphics and Applications*, 29(1):12–19, 2009.

[23] M. Chen and L. Floridi. An analysis of information in visualisation. *Synthese*, 190(16):3421–3438, 2013.

[24] M. Chen and A. Golan. What may visualization processes optimize? *IEEE Transactions on Visualization and Computer Graphics*, 2016. doi:10.1109/TVCG.2015.2513410.

[25] M. Chen and H. Jänicke. An information-theoretic framework for visualization. *IEEE Transactions on Visualization and Computer Graphics*, 16(6):1206–1215, 2010.

[26] M. Chen, S. Walton, K. Berger, J. Thiyagalingam, B. Duffy, H. Fang, C. Holloway, and A. E. Trefethen. Visual multiplexing. *Computer Graphics Forum*, 33(3):241–250, 2014.

[27] Marc Christie, Patrick Olivier, and Jean-Marie Normand. Camera control in computer graphics. *Computer Graphics Forum*, 27(8):2197–2218, 2008.

[28] M. C. Chuah and S. F. Roth. On the semantics of interactive visualizations. In *Proc. IEEE Information Visualization*, pages 29–36, 1996.

[29] Richard Cole and Howard Karloff. Fast algorithms for constructing maximum entropy summary trees. In Javier Esparza, Pierre Fraigniaud, Thore Husfeldt, and Elias Koutsoupias, editors, *Automata, Languages, and Programming*, volume 8572 of *Lecture Notes in Computer Science*, pages 332–343. Springer, Berlin, Heidelberg, 2014.

[30] Rodney Coleman. Markov chains. In *Stochastic Processes*, volume 14 of *Problem Solvers*, pages 35–61. Springer Netherlands, 1974.

[31] C. Correa and P. Lindstrom. The mutual information diagram for uncertainty visualization. *International Journal for Uncertainty Quantification*, 3(3):187–201, 2013.

[32] T. M. Cover and J. A. Thomas. *Elements of Information Theory*. John Wiley & Sons, Hoboken, NJ, 2nd edition, 2006.

[33] Thomas M. Cover and Joy A. Thomas. *Elements of Information Theory*. Wiley Series in Telecommunications. John Wiley & Sons, New York, 1991.

[34] Roger A. Crawfis and Nelson Max. Texture splats for 3D scalar and vector field visualization. In *Vis '93: Proceedings of the IEEE Visualization '93*, pages 261–266, 1993.

[35] James P. Crutchfield and David P. Feldman. Regularities unseen, randomness observed: Levels of entropy convergence. *Chaos*, 15:25–54, 2003.

[36] Imre Csiszár and Paul C. Shields. Information theory and statistics: A tutorial. *Foundations and Trends in Communications and Information Theory*, 1(4), 2004.

[37] A. Dasgupta, M. Chen, and R. Kosara. Measuring privacy and utility in privacy-preserving visualization. *Computer Graphics Forum*, 32(8):35–47, 2013.

[38] Aritra Dasgupta and Robert Kosara. Pargnostics: Screen-space metrics for parallel coordinates. *IEEE Transactions on Visualization and Computer Graphics*, 16(6):1017–1026, 2010.

[39] David Degani, Arnan Seginer, and Yuval Levy. Graphical visualization of vortical flows by means of helicity. *AIAA Journal*, 28(8):1347–1352, 1990.

[40] North Dakota State University Department of Mathematics. The mathematics genealogy project. http://www.genealogy.ams.org/.

[41] Michael R. Deweese and Markus Meister. How to measure the information gained from one symbol. *Network: Computation in Neural Systems*, 10(4):325–340, November 1999.

[42] B. Duffy, J. Thiyagalingam, S. Walton, D. J. Smith, A. Trefethen, J. C. Kirkman-Brown, E. A. Gaffney, and M. Chen. Glyph-based video visualization for semen analysis. *IEEE Transactions on Visualization and Computer Graphics*, 21(8):980–993, 2015.

[43] Mathias Eitz, Ronald Richter, Tamy Boubekeur, Kristian Hildebrand, and Marc Alexa. Sketch-based shape retrieval. *ACM Transactions on Graphics*, 31(4):31:1–31:10, July 2012.

[44] Miquel Feixas, Mateu Sbert, and Francisco González. A unified information-theoretic framework for viewpoint selection and mesh saliency. *ACM Transactions on Applied Perception*, 6(1):1:1–1:23, 2009.

[45] David P. Feldman. A brief introduction to: Information theory, excess entropy and computational mechanics. Lecture notes, Department of Physics, University of California, Berkeley (CA), USA, 2002. http://hornacek.coa.edu/dave/.

[46] R. A. Fisher. Theory of statistical estimation. *Mathematical Proceedings of the Cambridge Philosophical Society*, 22:700–725, 1925.

[47] Shachar Fleishman, Daniel Cohen-Or, and Dani Lischinski. Automatic camera placement for image-based modeling. *Computer Graphics Forum*, 19(2):101–110, 2000.

[48] J.-Y. Girard. Linear logic. *Theoretical Computer Science*, 50(1):1–102, 1987.

[49] A. Golan. *A Discrete Stochastic Model of Economic Production and a Model of Fluctuations in Production: Theory and Empirical Evidence.* PhD thesis, University of California, Berkeley, 1988.

[50] A. Golan. *Information and Entropy Econometrics: A Review and Synthesis*, volume 2 of *Foundations and Trends in Econometrics*. Now Publishers, 2008.

[51] A. Golan. On the foundations and philosophy of info-metrics. In *How the World Computes: Turing Centenary Conference and 8th Conference on Computability in Europe*, Springer LNCS 7318, pages 273–244. Springer, Berlin, Heidelberg, 2012.

[52] A. Golan, G. Judge, and D. Miller. *Maximum Entropy Econometrics: Robust Estimation with Limited Data.* John Wiley & Sons, Chichester, England, 1996.

[53] Francisco González, Miquel Feixas, and Mateu Sbert. View-based shape similarity using mutual information spheres. In *EG Short Papers*, pages 21–24, Prague, 2007. Eurographics Association.

[54] Francisco González, Mateu Sbert, and Miquel Feixas. Viewpoint-based ambient occlusion. *IEEE Computer Graphics and Applications*, 28(2):44–51, 2008.

[55] Amy Gooch, Bruce Gooch, Peter Shirley, and Elaine Cohen. A non-photorealistic lighting model for automatic technical illustration. In *Proceedings of the 25th Annual Conference on Computer Graphics and Interactive Techniques*, SIGGRAPH '98, pages 447–452, New York, New York, USA, 1998.

[56] S. Guiaşu. *Information Theory with Applications.* McGraw-Hill, New York, 1977.

[57] Martin Haidacher, Stefan Bruckner, and Eduard Groller. Volume analysis using multimodal surface similarity. *IEEE Transactions on Visualization and Computer Graphics*, 17(12):1969–1978, 2011.

[58] Martin Haidacher, Stefan Bruckner, Armin Kanitsar, and M. Eduard Gröller. Information-based transfer functions for multimodal visualization. In *Proceedings of the EG VCBM'08*, pages 101–108, 2008.

[59] R. W. Hamming. Error detecting and error correcting codes. *Bell System Technical Journal*, 29(2):147–160, 1950.

[60] R.V.L. Hartley. Transmission of information. *Bell System Technical Journal*, 7:535–563, 1928.

[61] Alfred Inselberg and B. Dimsdale. Parallel coordinates: A tool for visualizing multi-dimensional geometry. In *IEEE Visualization*, pages 361–378, 1990.

[62] Andrey Iones, Anton Krupkin, Mateu Sbert, and Sergey Zhukov. Fast, realistic lighting for video games. *IEEE Computer Graphics and Applications*, 23(3):54–64, 2003.

[63] H. Jänicke and M. Chen. A salience-based quality metric for visualization. *Computer Graphics Forum*, 29(3):1183–1192, 2010.

[64] H. Jänicke and G. Scheuermann. Visual analysis of flow features using information theory. *IEEE Computer Graphics and Applications*, 30(1):40–49, 2010.

[65] H. Jänicke, T. Weidner, D. Chung, R. S. Laramee, P. Townsend, and M. Chen. Visual reconstructability as a quality metric for flow visualization. *Computer Graphics Forum*, 30(3):781–790, 2011.

[66] Jinhee Jeong and Fazle Hussain. On the identification of a vortex. *Journal of Fluid Mechanics*, 285:69–94, 1995.

[67] Stefan Jeschke, David Cline, and Peter Wonka. A GPU Laplacian solver for diffusion curves and Poisson image editing. *ACM Transactions on Graphics*, 28(5):1–8, 2009.

[68] Guangfeng Ji and Han-Wei Shen. Dynamic view selection for time-varying volumes. *IEEE Transactions on Visualization and Computer Graphics*, 12(5):1109–1116, 2006.

[69] Bruno Jobard and Wilfrid Lefer. Creating evenly-spaced streamlines of arbitrary density. In *Proceedings of Eighth Eurographics Workshop on Visualization in Scientific Computing*, pages 43–55, 1997.

[70] Howard Karloff and Kenneth E. Shirley. Maximum entropy summary trees. *Computer Graphics Forum (Proc. EuroVis)*, 32(3):71–80, 2013.

[71] D. Keim, G. Andrienko, J. D. Fekete, C. Görg, J. Kohlhammer, and G. Melanon. Visual analytics: Definition, process, and challenges. In *Information Visualization: Human-Centered Issues and Perspectives*, Springer LNCS 4950, pages 154–175, 2008.

[72] G. Kindlmann and C.-F. Westin. Diffusion tensor visualization with glyph packing. *IEEE Transactions on Visualization and Computer Graphics*, 12(5):1329–1336, 2006.

[73] Gordon Kindlmann, Ross Whitaker, Tolga Tasdizen, and Torsten Moller. Curvature-based transfer functions for direct volume rendering: Methods and applications. In *Proceedings of the 14th IEEE Visualization 2003 (VIS'03)*, VIS '03, pages 513–520, Washington, DC, USA, 2003. IEEE Computer Society.

[74] Solomon Kullback. *Information Theory and Statistics (Dover Books on Mathematics)*. Dover Publications, Mineola, NY, 1997.

[75] Hayden Landis. Production-ready global illumination. In *SIGGRAPH 2002 Course 16*, pages 87–101, 2002.

[76] Teng-Yok Lee, Oleg Mishchenko, Han-Wei Shen, and Roger Crawfis. View point evaluation and streamline filtering for flow visualization. In *IEEE Pacific Visualization Symposium, PacificVis 2011*, pages 83–90, Hong Kong, China, March 2011.

[77] P. Legg, D. Chung, M. Parry, M. Jones, R. Long, I. Griffiths, and M. Chen. MatchPad: Interactive glyph-based visualization for real-time sports performance analysis. *Computer Graphics Forum*, 31(3):1255–1264, 2012.

[78] P. A. Legg, E. Maguire, S. Walton, and M. Chen. Glyph visualization: A fail-safe design scheme based on quasi-Hamming distances. *IEEE Computer Graphics & Applications*, to appear, 2016.

[79] P. A. Legg, E. Maguire, S. Walton, and Min Chen. Quasi-Hamming distances: An overarching concept for measuring glyph similarity. In *Computer Graphics and Visual Computing: Extended Abstract*, 2015.

[80] Bo Li, Yijuan Lu, and Henry Johan. Sketch-based 3D model retrieval by viewpoint entropy-based adaptive view clustering. In *Proceedings of the Sixth Eurographics Workshop on 3D Object Retrieval (3DOR '13)*, pages 49–56. Eurographics Association, Aire-la-Ville, Switzerland, 2013.

[81] Kewei Lu, Abon Chaudhuri, Teng-Yok Lee, Han-Wei Shen, and Pak Chung Wong. Exploring vector fields with distribution-based streamline analysis. In *IEEE Pacific Visualization Symposium, PacificVis 2013*, pages 257–264, Sydney, NSW, Australia, February 2013.

[82] E. Maguire, P. Rocca-Serra, S.-A. Sansone, J. Davies, and M. Chen. Taxonomy-based glyph design with a case study on visualizing workflows of biological experiments. *IEEE Transactions on Visualization and Computer Graphics*, 18(12):2603–2612, 2012.

[83] Nikolaos A. Massios and Robert B. Fisher. A best next view selection algorithm incorporating a quality criterion. In *Proceedings of the British Machine Vision Conference*, pages 78.1–78.10, Southampton, England, 1998. BMVA Press.

[84] A. Mebarki, P. Alliez, and O. Devillers. Farthest point seeding for efficient placement of streamlines. In *Visualization, 2005. VIS 05. IEEE*, pages 479–486, Oct 2005.

[85] Marina Meila. Comparing clusterings by the variation of information. In Bernhard Schölkopf and Manfred K. Warmuth, editors, *Learning Theory and Kernel Machines*, volume 2777 of *Lecture Notes in Computer Science*, pages 173–187. Springer, Berlin, Heidelberg, 2003.

[86] Konrad Mühler, Mathias Neugebauer, Christian Tietjen, and Bernhard Preim. Viewpoint selection for intervention planning. In *Proceedings of Eurographics/ IEEE-VGTC Symposium on Visualization*, pages 267–274, 2007.

[87] T. Munzner. A nested model for visualization design and validation. *IEEE Transactions on Visualization and Computer Graphics*, 15(6):921–928, 2009.

[88] H. Nyquist. Certain factors affecting telegraph speed. *Bell System Technical Journal*, 3:324–346, 1924.

[89] D. Oelke, D. Spretke, A. Stoffel, and D. A. Keim. Visual readability analysis: How to make your writings easier to read. In *Proc. IEEE VAST*, pages 123–130, 2010.

[90] Alexandrina Orzan, Adrien Bousseau, Holger Winnemöller, Pascal Barla, Joëlle Thollot, and David Salesin. Diffusion curves: A vector representation for smooth-shaded images. *ACM Transactions on Graphics*, 27(3):1–8, 2008.

[91] Maya Ozaki, Like Gobeawan, Shinya Kitaoka, Hirofumi Hamazaki, Yoshifumi Kitamura, and Robert W. Lindeman. Camera movement for chasing a subject with unknown behavior based on real-time viewpoint goodness evaluation. *The Visual Computer*, 26(6–8):629–638, 2010.

[92] Luis M. Portela. *Identification and Characterization of Vortices in the Turbulent Boundary Layer. Volume I.* PhD thesis, Stanford University, 1997.

[93] Helen C. Purchase, Natalia Andrienko, T.J. Jankun-Kelly, and Matthew Ward. Theoretical foundations of information visualization. In Andreas Kerren, JohnT. Stasko, Jean-Daniel Fekete, and Chris North, editors, *Information Visualization*, volume 4950 of *Lecture Notes in Computer Science*, pages 46–64. Springer, Berlin, Heidelberg, 2008.

[94] N. Rashevsky. Life, information theory, and topology. *Bulletin of Mathematical Biophysics*, 17:229–235, 1955.

[95] P. Rheingans and C. Landreth. Perceptual principles for effective visualizations. In Georges Grinstein and Haim Levkowitz, editors, *Perceptual Issues in Visualization*, pages 59–74. Springer-Verlag, 1995.

[96] Y. Rubner, C. Tomasi, and L.J. Guibas. A metric for distributions with applications to image databases. In *ICCV '98: Proceedings of the International Conference on Computer Vision*, pages 59–66, 1998.

[97] M. Ruiz, I. Boada, I. Viola, S. Bruckner, M. Feixas, and M. Sbert. Obscurance-based volume rendering framework. In Hans-Christian Hege, David Laidlaw, Renato Pajarola, and Oliver Staadt, editors, *IEEE/ EG Symposium on Volume and Point-Based Graphics*. The Eurographics Association, 2008.

[98] Marc Ruiz, Anton Bardera, Imma Boada, Ivan Viola, Miquel Feixas, and Mateu Sbert. Automatic transfer functions based on informational divergence. *IEEE Transactions on Visualization and Computer Graphics*, 17(12):1932–1941, 2011.

[99] Marc Ruiz, Imma Boada, Miquel Feixas, and Mateu Sbert. Viewpoint information channel for illustrative volume rendering. *Computers & Graphics*, 34(4):351–360, 2010.

[100] Marc Ruiz, Ivan Viola, Imma Boada, Stefan Bruckner, Miquel Feixas, and Mateu Sbert. Similarity-based exploded views. In *Proceedings of Smart Graphics'08*, pages 154–165, 2008.

[101] N. Sahasrabudhe, J.E. West, R. Machiraju, and M. Janus. Structured spatial domain image and data comparison metrics. In *Proc. IEEE Visualization*, pages 97–515, 1999.

[102] Alper Sarikaya, Danielle Albers, Julie C. Mitchell, and Michael Gleicher. Visualizing validation of protein surface classifiers. *Computer Graphics Forum*, 33(3):171–180, June 2014.

[103] Thomas Schreiber. Measuring information transfer. *Physical Review Letters*, 85:461–464, July 2000.

[104] David W. Scott. On optimal and data-based histograms. *Biometrika*, 66(3):605–610, 1979.

[105] Ekrem Serin, Serdar Hasan Adali, and Selim Balcisoy. Automatic path generation for terrain navigation. *Computers & Graphics*, 36(8):1013–1024, December 2012.

[106] C. E. Shannon and W. Weaver. *The Mathematical Theory of Communication*. University of Illinois Press, 1949.

[107] Claude E. Shannon. A mathematical theory of communication. *The Bell System Technical Journal*, 27:379–423, 623–656, July, October 1948.

[108] R. Shepard. Attention and the metric structure of the stimulus space. *Journal of Mathematical Psychology*, 1(1):54–141, 1964.

[109] B. Shneiderman. The eyes have it: A task by data type taxonomy for information visualizations. In *Proc. IEEE Symposium on Visual Languages*, pages 336–343, 1996.

[110] Noam Slonim and Naftali Tishby. Agglomerative information bottleneck. In *Proceedings of NIPS-12 (Neural Information Processing Systems)*, pages 617–623. MIT Press, 2000.

[111] Noam Slonim and Naftali Tishby. Document clustering using word clusters via the information bottleneck method. In *Proceedings of the 23rd Annual International ACM SIGIR Conference on Research and Development in Information Retrieval*, pages 208–215. ACM Press, 2000.

[112] Dmitry Sokolov, Dimitri Plemenos, and Karim Tamine. Methods and data structures for virtual world exploration. *The Visual Computer*, 22(7):506–516, 2006.

[113] W. Stallings. *Wireless Communications and Networks*. Pearson, Upper Saddle River, NJ, 2nd edition, 2004.

[114] H. P. E. Stern and S. A. Mahmoud. *Communication System: Analysis and Design*. Prentice-Hall, Upper Saddle River, NJ, 2003.

[115] A. James Stewart. Vicinity shading for enhanced perception of volumetric data. In *VIS 2003. IEEE Visualization, 2003*, pages 355–362, 2003.

[116] Alexander Strehl and Joydeep Ghosh. Cluster ensembles: A knowledge reuse framework for combining multiple partitions. *Journal of Machine Learning Research*, 3:583–617, March 2003.

[117] Shigeo Takahashi, Issei Fujishiro, Yuriko Takeshima, and Tomoyuki Nishita. A feature-driven approach to locating optimal viewpoints for volume visualization. In *IEEE Visualization 2005*, pages 495–502, 2005.

[118] G. K. L. Tam, H. Fang, A. J. Aubrey, P. W. Grant, P. L. Rosin, D. Marshall, and M. Chen. Visualization of time-series data in parameter space for understanding facial dynamics. *Computer Graphics Forum*, 30(3):901–910, 2011.

[119] Jun Tao, Jun Ma, Chaoli Wang, and Ching-Kuang Shene. A unified approach to streamline selection and viewpoint selection for 3D flow visualization. *IEEE Transactions on Visualization and Computer Graphics*, 19(3):393–406, March 2013.

[120] K. E. Taylor. Summarizing multiple aspects of model performance in a single diagram. *Journal of Geophysical Research*, 106:7183–7192, April 2001.

[121] J. Thiyagalingam, S. Walton, B. Duffy, A. Trefethen, and M. Chen. Complexity plots. *Computer Graphics Forum*, 32(3pt1):111–120, 2013.

[122] J. J. Thomas and K. A. Cook. *Illuminating the Path: The Research and Development Agenda for Visual Analytics*. National Visualization and Analytics Center, 2005.

[123] Naftali Tishby, Fernando C. Pereira, and William Bialek. The information bottleneck method. In *Proceedings of the 37th Annual Allerton Conference on Communication, Control and Computing*, pages 368–377, 1999.

[124] D. G. Tucker. The early history of amplitude modulation, sidebands and frequency-division-multiplex. *Radio and Electronic Engineer*, 41(1), 1971.

[125] Lisa Tweedie. Characterizing interactive externalizations. In *Proceedings of the ACM SIGCHI Conference on Human Factors in Computing Systems*, CHI '97, pages 375–382, New York, NY, USA, 1997. ACM.

[126] UCIrvine. Machine learning repository. `http://archive.ics.uci.edu/ml/`.

[127] C. Upson, T. Faulhaber, Jr., D. Kamins, D. H. Laidlaw, D. Schlegel, J. Vroom, R. Gurwitz, and A. van Dam. The application visualization system: A computational environment for scientific visualization. *IEEE Computer Graphics and Applications*, 9(4):30–42, 1989.

[128] M. J. Usher. *Information Theory for Information Technologists*. MacMillan, Basingstoke, England, 1984.

[129] J. J. van Wijk. The value of visualization. In *Proc. IEEE Visualization*, pages 79–86, 2005.

[130] Pere P. Vázquez, Miquel Feixas, Mateu Sbert, and Wolfgang Heidrich. Viewpoint selection using viewpoint entropy. In Thomas Ertl, Bernd Girod, Gerhard Greiner, Heinrich Niemann, and Hans-Peter Seidel, editors, *Proceedings of Vision, Modeling, and Visualization 2001*, pages 273–280, Stuttgart, Germany, November 2001.

[131] Pere P. Vázquez, Miquel Feixas, Mateu Sbert, and Antoni Llobet. Real-time automatic selection of good molecular views. *Computers & Graphics*, 30(1):98–110, 2006.

[132] Pere-Pau Vázquez, Miquel Feixas, Mateu Sbert, and Wolfgang Heidrich. Automatic view selection using viewpoint entropy and its application to image-based modelling. *Computer Graphics Forum*, 22(4):689–700, 2003.

[133] Pere-Pau Vázquez and Mateu Sbert. Fast adaptive selection of best views. In Vipin Kumar, Marina L. Gavrilova, Chih Jeng Kenneth Tan, and Pierre L'Ecuyer, editors, *Computational Science and Its Applications — ICCSA 2003*, volume 2669 of *Lecture Notes in Computer Science*, pages 295–305. Springer, Berlin, Heidelberg, 2003.

[134] Sergio Verdú. Fifty years of Shannon theory. *IEEE Transactions on Information Theory*, 44(6):2057–2078, October 1998.

[135] Vivek Verma, David T. Kao, and Alex Pang. A flow-guided streamline seeding strategy. In *Vis '00: Proceedings of the IEEE Visualization 2000*, pages 163–170, 2000.

[136] Ivan Viola, Miquel Feixas, Mateu Sbert, and M. Eduard Gröller. Importance-driven focus of attention. *IEEE Transactions on Visualization and Computer Graphics*, 12(5):933–940, 2006.

[137] Chaoli Wang and Han-Wei Shen. LOD Map: A visual interface for navigating multiresolution volume visualization. *IEEE Transactions on Visualization and Computer Graphics*, 12(5):1029–1036, 2006.

[138] Chaoli Wang, Hongfeng Yu, Ray W. Grout, Kwan-Liu Ma, and Jacqueline H. Chen. Analyzing information transfer in time-varying multivariate data. In *Proceedings of IEEE Pacific Visualization Symposium 2011*, pages 99–106, 2011.

[139] Chaoli Wang, Hongfeng Yu, and Kwan-Liu Ma. Importance-driven time-varying data visualization. *IEEE Transactions on Visualization and Computer Graphics*, 14(6):1547–1554, 2008.

[140] M. O. Ward. A taxonomy of glyph placement strategies for multidimensional data visualization. *Information Visualization*, 1(3/4):194–210, 2002.

[141] M. O. Ward and J. Yang. Interaction spaces in data and information visualization. In *Proc. Eurographics/IEEE TCVG Symposium on Visualization*, pages 137–145, 2004.

[142] Tzu-Hsuan Wei, Teng-Yok Lee, and Han-Wei Shen. Evaluating isosurfaces with level-set-based information maps. *Computer Graphics Forum*, 32(3):01–10, 2013.

[143] N. Wiener. *Cybernetics: Or Control and Communication in the Animal and the Machine*. John Wiley & Sons, New York, 1948.

[144] P. C. Wong and J. Thomas. Visual analytics. *IEEE Computer Graphics and Applications*, 24(5):20–21, 2004.

[145] Chenyang Xu and Jerry L. Prince. Gradient vector flow: A new external force for snakes. In *CVPR '97: Proceedings of the IEEE Computer Vision and Pattern Recognition 1997*, page 66, 1997.

[146] Lijie Xu, Teng-Yok Lee, and Han-Wei Shen. An information-theoretic framework for flow visualization. *IEEE Transactions on Visualization and Computer Graphics*, 16(6):1216–1224, 2010.

[147] J. Yang-Peláez and W. C. Flowers. Information content measures of visual displays. In *Proceedings of the IEEE Symposium on Information Vizualization 2000*, pages 99–103, 2010.

[148] Raymond W. Yeung. *Information Theory and Network Coding*. Information Technology: Transmission, Processing and Storage. Springer, New York, US, 2008.

[149] J. S. Yi, Y. ah Kang, J. T. Stasko, and J. A. Jacko. Toward a deeper understanding of the role of interaction in information visualization. *IEEE Transactions on Visualization and Computer Graphics*, 13(6):1224–1231, 2007.

[150] H. Zhou, M. Chen, and M.F. Webster. Comparative evaluation of visualization and experimental results using image comparison metrics. In *Proceedings of IEEE Visualization*, pages 315–322, 2002.

Index

ACR, *see* alphabet compression ratio
alphabet, **18**, 19, 24–26, 29, 34–43,
      46–48, 58, 60
    data alphabet, 18, 23, 25, 26, 36,
        37, 47
    graphics alphabet, 19
    transformation, 34, 35, 37–40,
        43, 47, 48
alphabet compression ratio, 37
ambient occlusion, 66, 80, 82, 83
analytical visualization, 42, 47
AO, 80, 82–86
application
    document readability, 47
    facial expression, 48
    file system, 62
    sports, 46
    video classification, 48

binary digits view, 19, 20
binning, 120
bit, 18, 39, 61
block, 97

camera path, 91
causal relationship, 164
CBR, *see* cost–benefit ratio
channel, *see* communication channel
channel coder, 26, 27
classification, 121
clipping, 104
color ambient occlusion, 74, 86
communication channel, 1, 22, 24–25,
    117, 119, 151
comparative visualization, 54
conditional entropy, 3, 29, 30, 90,
    103, 160
contribution, 96
correlation, 171

cost–benefit, 21, **32–48**
cost–benefit ratio, 39, 41
cumulative similarity, 100

D_{KL}, *see* Kullback–Leibler distance
data object, 17
data processing inequality, 6, 19, **34**,
    37, 42
decision tree, 48
decomposition of mutual
    information, 118
demultiplexer, 49, *see also* visual
    multiplexing
demux, 49, *see also* visual
    multiplexing
dendogram, 162
display space capacity, 19, 55, 56
display space utilization, 20–22,
    26–27, 29, 55, 56
disseminative visualization, 41, 45
dissimilarity, 106
distance volume, 99, 102, 120
distortion, 96
DSU, *see* display space utilization
dynamic programming, 91
dynamic view selection, 91

ECR, *see* effectual compression ratio
effectual compression ratio, 38
entanglement, 118
entropy, 2, 18–22, 25, 26, 29, 30,
    34–39, 41, 45, 46, 56, 57,
    93, 97, 99, 156, 158, 174
entropy density, 8
entropy encoding, 26
entropy rate, 8, 9, 105
entropy-based spatial mapping, 26
error correction, 27–29, 58–63
error detection, 27–29, 58–63